To Iain October 2001 —
With many thanks for all
your caring conservation ideas
Much love from Janey

Caring for the past

To Nick

Caring for the past

Issues in conservation for archaeology and museums

Elizabeth Pye

Published by James & James (Science Publishers) Ltd,
35–37 William Road, London NW1 3ER, UK

A catalogue record for this book is available from the British Library.

ISBN 1 902916 10 7

Designed and Typeset by Hardlines, Charlbury
Printed in the UK by Hobbs the printers

The illustrations on the front cover are:

Eighteenth century houses inserted into the ruins of the abbey at Bury St Edmunds, Suffolk, UK. Courtesy of N D Balaam.

Eighteenth century leather fire bucket after restoration. Courtesy of Royal & Sun Alliance Insurance Group plc and the Leather Conservation Centre.

Removal of superficial 'dirt' from a Roman Mummy portrait, clarifying the flesh tones and the modelling of the face, Petrie Museum of Egyptian Archaeology, UCL, UC 14692. Courtesy of H & R Jaeschke.

Greek Lexicon published in 1738 and covered with schoolboy scribbles in 1834. Courtesy of N D Balaam.

Saxon brooch after removal of corrosion products. Courtesy of UCL Institute of Archaeology.

Museum stores in Zimbabwe after reorganization by museum staff during a short training course organized by PREMA. Courtesy of Gaël de Guichen.

Contents

Preface

This book was formed during a sabbatical year in 1998–9 and was completed at irregular intervals during the following year. It draws on many of the ideas I have developed over the period during which I have been working in conservation, and it is influenced by current issues. Over the two years when I was writing it my ideas shifted and changed, but there were several concerns which have remained with me and are, to varying extents, visible in the book.

Conservation of the cultural heritage is a small discipline and the philosophy and practice within it is varied. I am interested in the common threads, and concerned that we should find a common voice. Conservators have fought hard for recognition and for better standards, but has this meant that we have not thought enough about why we are conserving, and for whom? Are we in danger of protecting objects from the very people for whom we are conserving them, rather than encouraging these same people to study and enjoy the objects (to use them), and to share in the process of their conservation?

Having come from a family of artists, designers and engineers, I have always been fascinated by how things are made and the way they look. Are we losing sight of (or downplaying) the fascination of the object, or failing to transmit this to other people? Have we, in some branches of conservation, so emphasized the importance of scientific evidence that we have lost sight of the importance of the visual impact of objects? Has the emphasis on the science of conservation, and increasingly on preventive conservation (both of which I support), resulted in the undervaluing of the practical and aesthetic skills needed for remedial conservation and restoration?

I have not wanted these concerns to dominate the book, but they surface from time to time. The finished product is an attempt to look at the current issues relating to the conservation of archaeological and museum objects, and to see them in the wider heritage context. It is not a book about how to conserve objects; rather it is intended to stimulate discussion and debate. Because I have tried to look for common ground between different fields of conservation, my examples are not limited to museums, or what is normally seen as archaeology.

The book is deliberately non-technical. It assumes a broad operational understanding of conservation and is intended for anyone who wants to explore ideas about conservation of archaeological and museum objects, including heritage or collections managers, archaeologists, museum curators and materials scientists. It is also intended for conservators – either those embarking on their training or those in mid-career who wish to reflect on the way conservation is developing, and to share in the debate about the role it should play in the future.

Elizabeth Pye
Summerfield, Norfolk
September 2000

Acknowledgements

I have many people to thank. This book would not have happened without a sabbatical year, and for that I must thank Peter Ucko wholeheartedly (and for his continued insistence that I actually finish the book). I must also thank my colleagues at University College London Institute of Archaeology who were very tolerant of my absences, even when I was no longer on sabbatical.

The book was shaped not just by the reading and thinking which I did during the last two years but by many earlier experiences, and by the ideas and friendship of many people I have worked with. There are too many to mention all by name, but I want, especially, to mention the late Nigel Williams, with whom I worked at the British Museum when I was first finding my feet in conservation; his skills, ingenuity and craftsmanship were impressive. I would also like to express my special thanks to all the students I have worked with. I hesitate to say 'taught' because I have learned a great deal from them, especially as they have always had the knack of asking the one question I had not anticipated! Another important influence has been my long-term links with ICCROM (the International Centre for Conservation, Rome). I would particularly like to thank all the friends and colleagues I have shared ideas with, and learned from, during the PREMA programme (Training for Conservation Management in Sub-Saharan Africa, initiated by ICCROM in 1987), especially Catherine Antomarchi, Bob Ferguson, Gaël de Guichen, Baba Keita, Kwame Labi, Terry Little, Mubiana Luhila and Margaret McCord.

In producing the book, I am very grateful indeed for the continuing support and friendship of Janey Cronyn and Kate Foley, who encouraged me from start to finish, discussed ideas and read parts of the text. I particularly thank Elaine Sansom, who read the whole text and provided many pertinent comments.

I would also like to thank very warmly all the following who (in spite of the pressures in their own lives) willingly gave me information, lent me books and articles or criticized parts of what I had written: Laura Drysdale, Jane Hubert, Susanne Keene, John Merkel, Nick Merriman, Gustav Milne, Clifford Price, Carolyn Rose, Tim Schadla-Hall, Nigel Seeley, Nicholas Stanley-Price, Kate Starling, Kirsten Suenson-Taylor, Dean Sully, Peter Ucko and Glenn Wharton.

I would also like to thank equally warmly all those who helped with illustrations, or gave me permission to use their own photographs: Nick Balaam, Angela Evans, Gaël de Guichen, Helena and Richard Jaeschke, Stuart Laidlaw, Gustav Milne, Francis Pryor, Theo Sturge, Kathy Tubb, and the staff of the Conservation Department at the Museum of London, particularly Liz Barham, Andrew Calver, Helen Ganiaris, Claire Heywood and Rob Payton.

Finally, I would like to record my affectionate thanks to my family, Nick, Fran and David, who have endured an invasion of books and paper, and an often absentee wife and mother. Nick applied logic and structure whenever my thoughts became too rambling, or my prose too purple, and saw me through the final stages of production quite magnificently: the book owes an enormous amount to his support.

1 Archaeology, museums and the 'heritage industry': a context for conservation

Introduction

This book is about conservation of objects and structures relating to the past, often referred to as the material heritage. Many of these objects are described as 'archaeological' and archaeology is increasingly widely interpreted as including material relating to the recent, as well as the distant past, and relating to other cultures. The principal focus of this book is on archaeological materials which have been revealed through excavation, many of which are subsequently placed in museums. It also refers to the wide range of objects in museums which not only relate to the past but to a variety of disciplines such as art history, social history or ethnography.

This chapter introduces the context in which conservation now takes place by discussing the changes which have occurred in archaeology, museums and the private sector. Some of what follows will be familiar to some of the readers of this book but the aim is to bring together the changes and briefly review their effects on conservation. Some of these changes are influenced by the increasing public interest in all forms of the past which has led to the growth of a wide-ranging phenomenon known as 'heritage', or sometimes as the 'heritage industry' (Hewison 1987). Many of the current issues in archaeology and museums arise from the opening-up of their disciplines and practices, as there has been a shift away from the old certainties of 'fact' and 'truth' toward multiple and varied interpretations. These issues have a direct impact on what is expected of conservation and on how conservation itself is developing.

Introduction to conservation

Conservation is the action of safeguarding the objects and structures which compose the material remains of the past, and it aims to ensure that these remains are available to use and enjoy today and in the future. In its widest sense it covers the work of government bodies such as English Heritage, or the National Parks Service in the USA. In protection and conservation of the cultural heritage in a narrower sense, it also covers the structural and other changes which may be made to a building or object to repair damage or prevent deterioration. Therefore it embraces a range of activities from protective legislation to laboratory treatments. The term conservation is not universally applied (eg 'Historic Preservation' is used in the USA to denote conservation of buildings and monuments). These semantic complications are discussed further in Chapter 2.

This book focuses on conservation in the narrower sense of repairing or otherwise treating objects themselves. This form of conservation is a relatively young applied science which has grown out of earlier craft traditions of restoration and now involves the theory and practice of technology and deterioration studies, and of conservation procedures. It is closely linked to, or is part of, disciplines such as archaeology and art history which involve the study of the heritage (Fig 1), but it is recognized as a distinct field with its own codes of ethics and professional associations.

Figure 1 Audley End House, Essex, UK. The picture gallery, showing a range of different aspects of the cultural (and natural) heritage, including the building itself (Copyright English Heritage).

Conservation procedures are applied after an object or structure has been made and used (and usually after it has been taken into a museum or other collection, or has been designated as being significant). Conservation procedures are used to assess the nature and condition of objects, to rectify or ameliorate the effects of poor manufacture, of use and decay, and to prevent or minimize the processes of deterioration on site or in the museum. Some conservation procedures are used to prepare objects or structures for an interpretative role in museums or on site. 'Conservation is a process that consistently recreates its product (cultural heritage)' (Avrami *et al* 2000b, 7), thus it is important that it is aware of, and articulates with, the context in which it takes place.

Where conservators work

Today conservators are employed in institutions such as museums, universities and government bodies concerned with the heritage; they also work independently on single or multiple projects for public or private bodies, antiques dealers and private individuals. Conservators employed by large institutions may be specialist members of a conservation team, and may have access to a wide range of scientific advice and facilities. By contrast, in smaller institutions such as local or regional museums, there may be a single conservator who has responsibility for the care of a wide-ranging collection, or there may be no conservator on the staff, the conservation being carried out by independent conservators when the need arises. Some independent conservators work almost entirely for the antiques trade, and for private individuals, conserving 'collectable' objects such as furniture and paintings.

The changing context of conservation

Archaeology

The discipline of archaeology emerged during the nineteenth century and was concerned at first with the identification of the material remains which provided evidence of a distant past, and with proposing new chronologies – initially contentious since they overturned the previous orthodoxies (Daniel 1975, Schnapp 1996). It developed as a descriptive discipline: attempting to reveal what had happened in the past through examination and classification of structures and artefacts. Techniques of excavation were developed and refined, technologies such as photography and surveying were harnessed for the recording of the material remains and conservation procedures were developed. New methods and approaches were debated and proposed during the second half of the twentieth century, particularly in Anglo-American archaeology, in which the processes involved in the creation of the archaeological record became the primary focus of study. Archaeology now sought to be seen as objective and scientific, and to explain the past in terms of natural and social processes: scientific techniques of physical and chemical analysis were increasingly adopted to investigate material remains. In the 1980s there was a reaction (known as post-processual archaeology) which appeared to turn away from the scientific and adopt a more subjective approach. During the 1990s the debate shifted from the polarization of post-processual versus processual archaeology to focus on the ways in which views of the past were affected by the politics and ideologies of the present (see the volumes in the series One World Archaeology [OWA], such as Layton 1989a, Layton 1989b). One development from this was the concept of 'interpretative archaeology' (Shanks 1992, Tilley 1993, Hodder *et al* 1995).

Different interpretations of the past

Nowadays the archaeological debate has opened up to embrace different interpretations of the past, and thus acknowledge the existence of many different archaeologies (and different influences including the sciences, sociology, art history and anthropology). The emphasis is on 'making sense' of the past while accepting that it is never possible to 'know' the past – it will always be uncertain and open to different interpretations which are affected by the context (eg cultural, political) in which they take place. The past is shaped, and reshaped, in the present, just as musicians interpret and reinterpret musical scores, and actors and directors endlessly reinterpret Shakespeare's plays, different interpreters bringing out different meanings (Tilley 1993, Hodder *et al* 1995).

Widening audiences

The acceptance that many different interpretations of the past are possible has led to the acknowledgement that the past is not the exclusive preserve of experts (ie professional archaeologists), and is beginning to open archaeology to wider audiences. An increasingly important audience, certainly in the UK, is school students influenced by developments in the National Curriculum; there is also a growing general audience whose interest is sparked by television programmes about archaeology.

Some archaeologists are moving away from the academic mode of debate and presentation to a much more inclusive approach, which takes account of the views of groups who have hitherto been marginalized, such as indigenous peoples (Layton 1989a, Layton 1989b) or people who adopt an alternative life-style such as 'free festivalers' (Bender 1998). The World Archaeological Congress (WAC) has pioneered meetings which include both academics and non-academics:

> *Many of the latter (ie non-academics), accustomed to being treated as the 'subjects' of archaeological and anthropological observation, had never before been admitted as equal participants in the discussion of their own (cultural) past or present, with their own particularly vital contribution to make towards global, cross-cultural understanding. (Ucko 1989, ix)*

Listening to different views has also led to re-examination of archaeological practice particularly influenced by the beliefs and sensitivities of people most concerned with material heritage. For example, in 1989 the World Archaeological Congress Inter-Congress meeting at Vermillion supported the *Vermillion Accord*, which sets out ways in which human remains should be respected. In 1990 WAC adopted the First Code of Ethics' setting out principles which arise from obligations to indigenous peoples (Hubert 1989a, WAC 1991). In 1990 the *Native American Graves Protection and Repatriation Act* was passed, which provides a legislative framework for the return of human remains and grave-goods from museums to Native American Groups (NAGPRA 1990).

Archaeological interpretation remains dependent on material remains and archaeology can be described as 'the analysis of past material cultures' (Ucko 1995, 2) or the analysis of the material heritage. But the archaeological past extends right up to yesterday, and today's material culture will be tomorrow's heritage, or the stuff of tomorrow's archaeological analysis. Therefore 'archaeology' can embrace recent events, and the material heritage of widely differing cultures. It is no longer concerned with disinterested knowledge and analytical study of material remains, and has been transformed into a wide-ranging discipline which engages with society today.

Archaeological conservation

Archaeological conservation is the branch of conservation concerned with the objects and structures which constitute the archaeological record, and particularly with the problems of excavated materials. Because much excavated material is only partially preserved, archaeological conservation is sometimes likened to forensic science in that establishing the nature of the object may require similarly detailed investigation of often tiny and deteriorated traces of material which provide the clues to identity and significance (eg traces of leather and wood indicating the original presence and structure of a scabbard on an iron sword).

This investigation phase reveals information on raw materials, technology and function and it is followed by conservation procedures which make the objects and structures strong and stable enough to survive into the future (and thus be available for further research). The investigation of excavated artefacts plays an important part in interpretation of the activities of the peoples associated with an excavated site: in this sense conservation is part of the discipline of archaeology (Cronyn 1990). Much of the information derived is primarily useful for archaeologists and other specialists, and only a very small proportion of excavated objects ends up on public display. However, with the widening of audiences the use of archaeological material has also widened. Some of the material, which hitherto was destined for long-term storage in archaeological archives, is now used in a number of new ways particularly in 'hands-on' activities, in museums, in schools, and in centres such as the Archaeological Resource Centre (ARC) at York, UK.

Museums

Many modern museums have their roots in the private collections built up in Europe from late mediaeval times onwards by royalty and the church, and by rich merchants and noblemen (such as the Medici dynasty in Florence). The Renaissance was characterized by an interest in antiquities and in the natural world, and collections tended to take the form of 'cabinets of curiosities' containing, for instance, paintings and sculpture as well as natural specimens such as stuffed animals, and sea-shells. There was no distinction between archaeological, natural history and fine art collections because these categories did not yet exist. Some collections were classified, and displayed in elaborately fitted rooms, but they were essentially for private gratification (Impey & MacGregor 1985, Hooper-Greenhill 1992, Lewis 1992). During this period damaged artefacts were restored by craftsmen who were otherwise normally involved in making similar objects; the emphasis was on maintaining the appearance of the object, and the concept of long-term preservation had not yet developed.

During the seventeenth century the idea of the public collection began to evolve; for example the collection presented to the University of Oxford by Elias Ashmole became the basis of the Ashmolean Museum which was opened to visitors (though presumably only a fairly narrow social range) in 1683. Public museums were founded in Europe and North America during the eighteenth and early nineteenth centuries (eg the British Museum in 1753, and the Smithsonian Institution in 1846), but the greatest growth world-wide occurred in the second half of the nineteenth century with museums and galleries created throughout Europe, North and South America, as well as in Africa, India, the far East and Australasia. Many, such as the National Museum of Antiquities in Copenhagen, were overtly national in character and were intended to promote a sense of national identity. During the late nineteenth and early twentieth centuries new types of museum began to emerge such as 'open-air' museums of buildings (the first example, Skansen, was opened in 1891 near Stockholm, Sweden) and museums of science and industry (Alexander 1983, Lewis 1992). Archaeological conservation first developed in museums; for example effective treatments for waterlogged wood and for corroded metals were devised during the second half of the nineteenth century in the National Museum of Antiquities in Copenhagen. Here the emphasis was not so much on the appearance of the object as on the evidence it contained (Brinch Madsen 1987). Scientific conservation was established with the appointment of a chemist to the Königlichen Museen, Berlin, in 1888 (Gilberg 1997).

New audiences

During the first half of the twentieth century museums tended to appeal to a narrow, well-educated and affluent audience, but from that time the growth of independent museums, including an increase in open-air and industrial museums, began the process of appealing to wider audiences. Now, at the beginning of the twenty-first century, most museums have become very different places. They have been working hard to lose their dusty reputation and to welcome visitors from all sections of the communities that they serve, largely by changing their emphasis from scholarship to education, and their image from exclusivity to accessibility. New displays and new museums such as the National Museum of Australia (to open in 2001) are attempting to be relevant to the whole of their community. 'If Australia's new museum is to play a role in national reconciliation, the politicians must restrain the celebratory impulse from overriding the indigenous curators' work to address Aboriginal history' (Leggett 1999, 17). In some museums the processes of conservation are themselves on display, appealing to audiences interested in the intricacies of laboratory technologies, or in how to care for their own possessions.

Museums today

There are many definitions of the modern museum but they share the same general concepts, the two key ones being that a museum should both conserve the collections it is responsible for, and enable them to be used. The use should involve both visitors (the enjoyment and inspiration of the public, and research by visiting scholars) and the staff of the museum itself (research by curators). In its statutes ICOM (the International Council for Museums) defines a museum as an institution which:

> *acquires, conserves, researches, communicates and exhibits, for purposes of study, education and enjoyment, material evidence of people and their environment. (ICOM 1996, Article 2)*

It goes on to say that other institutions or structures such as natural, archaeological, historical and ethnographic monuments and sites can be considered as museums. The ICOM definition is a very broad one and embraces anything which has the characteristics of a museum in that it communicates and displays material evidence of human activity.

The UK Museums Association puts more emphasis on the active involvement of the public but again the key activities are clearly to safeguard and to use the artefacts and specimens:

> *Museums enable people to explore collections for inspiration, learning and enjoyment. They are institutions that collect, safeguard and make accessible artefacts and specimens, which they hold in trust for society.* (Museums Journal *May 1999, 51)*

Thus museums, too, have opened up to using their collections in new ways to attract new audiences. Loans are made to schools and local communities, and in some cases material is repatriated to its original owners or their successors. Furthermore, in engaging in today's social issues, not only are collections used in different ways, but exhibitions may focus not just on objects but on other forms of evidence (sound recordings, film) to illustrate topics of current social interest.

Conservation in museums today

Conservation in museums has, in the past, focused almost entirely on making objects displayable, and has been largely exhibition-led, in that work concentrated on the objects required for the next exhibition. The overall care of collections and the conditions in which they were housed has been a much greater concern since the second half of the twentieth century, and in the last decade of that century the balance of responsibility shifted towards the care of whole collections (preventive conservation), including improving conditions for stored collections. Where the emphasis had been largely on the conservation of archaeological material and works of art, the effect of this growing focus on preventive care of whole collections was for conservation to engage with a much wider range of museum artefacts including, for example, natural history and social history material. Thus preventive conservation has broadened the responsibilities of museum conservators.

By the end of the twentieth century the opening up of archaeology and museums, and the pressures to make museums accessible, meant that objects were being used in new ways (involving more handling and more demonstration) with obvious effects on conservation. For conservators there have been changes in employment patterns. Some museums no longer employ in-house conservators, and many buy-in the services of independent conservators to deal with particular projects.

Heritage

The Oxford English Dictionary definition of heritage is 'that which has been or may be inherited; anything given or received to be a proper possession'. Heritage is normally seen as not just what has been inherited from an earlier generation but what people in the present have chosen to keep, both natural and manufactured, tangible and intangible. Thus the heritage can embrace rivers and mountains, language, music and traditions, objects and structures, any or all of which relate to the history and development of a society, and are considered worth preserving for that society and its future representatives. The significance of heritage may be personal or local, or far wider; thus heritage is often qualified, eg 'national heritage', or 'world heritage'. In this sense heritage has 'always' existed, but it has also been constructed and reconstructed in accordance with social changes (Pearce 2000).

The 'heritage industry'

The interest in heritage can also be seen as a (recent) social phenomenon involving different groups in using archaeological and historical artefacts, interpretations, memories and imagination to create and develop their own pasts or histories. For some, heritage and history (and archaeology) are part of a continuum, but both Merriman (1996) and Lowenthal (1997) pointed out that much of what is presented as heritage is not historically accurate, and Shanks saw heritage as being 'a sacrifice of the past for the present' and about 'what the past can do for the present' (Shanks 1992, 108). Lowenthal went on to say that this lack of accuracy is often deliberate: 'heritage' is intended to have impact because it is about selecting and transmitting particular messages often to do with forging identities.

> *Heritage the world over not only tolerates but thrives on and even requires historical error. Falsified legacies are integral to the exclusive purpose of group identity. (Lowenthal 1997, 132)*

He used examples such as the Swiss clinging to the story of William Tell even when his 'defiance of the Hapsburg oppressor is a notorious fiction' (op cit, 130) and the English unwillingness to discard the myth that Francis Drake played a major part in the defeat of the Spanish Armada.

McBryde pointed out that manipulation of the past is not new.

> *The philosophic issues of who owns, or controls, the present (or the future) by manipulations of perceptions of the past, or versions of the past, have always been vital. (McBryde 1985, 1)*

This manipulation can affect both the intangible and the tangible remains of the past, seen, for example, in the imposition of alien heritages on colonized peoples resulting in the obliteration of indigenous traditions (McBryde 1985, Layton 1989a, Layton 1989b, Lowenthal 1997), and in the nineteenth-century urge (in the UK) to strip and improve ancient buildings, particularly churches, to regain 'the purity they were believed to have had' (Lowenthal & Binney 1981, 20, Jokilehto 1999).

New audiences

Today's interest in 'heritage' epitomizes the inclusion of new and different groups seen in the development of archaeology and museums. Heritage is no longer the exclusive preserve of the 'elite', more people want to discover and use the past than ever before.

> *Heritage expands especially because more people now have a share in it. In times past, only a small minority sought forebears, amassed antiquities, enjoyed Old Masters, or toured museums and historic sites. (Lowenthal 1997, 10)*

Indigenous peoples who have been denied their past, and people whose 'ordinary' past has been seen as unimportant, want to claim their own identity and are now involved in reconstructing or discovering their heritages. This is seen in the increased interest in collecting past objects of all kinds, in genealogy and oral history, in the regeneration of traditional crafts and the preservation of vernacular buildings (Samuel 1994). For many the fascination of metal detecting must be just as much about being involved in rediscovering the lost past as about any hope of financial gain.

Attitudes to artefacts

That different groups may have different views of the past and its material remains is well illustrated by reactions to Stonehenge, the prehistoric stone circle and World Heritage Site in the south of England. The monument is the subject of multiple interpretations and is hotly disputed, with some groups concerned with protecting it (English Heritage, the National Trust, the police) in apparent, or actual, opposition to others wanting to touch it and use it in a number of different ways (tourists, druids, festivalers).

> *The designated custodians see it as something to be preserved, a museum piece, an artefact, a vital part of our rooted, stable, national identity. But for a small minority it is something different: a living site, a spiritual centre, an integral part of an alternative lifestyle. (Bender 1998, 9)*

The evolution of the phenomenon of heritage has meant that artefacts are both viewed and used in new ways. They may be searched for and collected, used more widely in education, or used as sources of information for reconstructing past heritages (eg indigenous peoples rediscovering their traditional technologies, or re-enactment groups researching details of costume of a particular period). Interest groups express strong feelings about their heritage (eg actors concerned with the preservation of the remains of Shakespeare's theatre, or a local community 'defending' a village against alteration or development). Conservation must now be more flexible and versatile in order to encompass new views and new uses.

The changing material heritage

The UNESCO *Convention Concerning the Protection of World Cultural and Natural Heritage* deals with sites and monuments, part of the material heritage. Articles 1 and 2 of this Convention define the material heritage as including architectural structures, monumental painting and sculpture; groups of buildings; sites made or modified by human action, including archaeological sites; and natural formations including biological and geological. If this description is scaled down the material heritage also includes tools and equipment, personal ornament, textiles, books, paintings, sculpture, botanical, zoological and geological specimens, as well as assemblages of any or all of these; in other words the kinds of objects or samples which might be excavated or collected, housed in a museum and so become subject to conservation measures, in the narrower sense (UNESCO 1972).

The composition of the material heritage is no more static than language or ritual; it is subject to re-evaluation and to addition, as well as to depletion through destruction or decay. The full extent of the buried and underwater heritage is unknown. It is not composed simply of known and designated monuments and sites, and objects excavated and placed in museums, it also embraces the archaeological deposits as yet undetected (Carver 1996). Re-evaluation of the material heritage is well-illustrated by attitudes to nineteenth-century, high Victorian buildings in the UK. Although by the middle of the twentieth century these buildings had become unfashionable and disregarded, and were likely to be demolished, by the end of that century they were seen as a valued part of the heritage and were largely protected. Plastic objects, mass-produced from the mid twentieth century and at first seen as cheap substitutes for natural materials, became valued and collected by the end of the century. Major wars during the twentieth century were responsible for the destruction of, or damage to, many valued buildings and collections. Familiar and commonplace objects such as domestic or small industrial equipment, or costume, are being broken, worn out and thrown out every day, long before they are perceived as heritage, thus increasing the ultimate heritage value of the few that remain. Although the material heritage is usually described as a non-renewable resource, some have argued that new discoveries and new additions mean that it is constantly renewed (Holtorf and Schadla-Hall 1999).

Protecting the heritage

Lowenthal points out that the concept of heritage 'offers a rationale for self-respecting stewardship of all we hold dear' (Lowenthal 1997, 2). Although things passively inherited may not be valued now (though they may have been in the past and may be again), things deliberately chosen are valued by the choosers, and 'the heritage' is generally agreed to represent things which are considered significant enough to be worth protecting for present and future use, for enjoyment, or to commemorate events such as the holocaust, ie for 'making sense of misery' (Thomas 1999, 30).

Protection for the future implies that heritage is held in trust and should be cared for to ensure that it survives to be handed on. This is not a new idea; for example in the nineteenth century Ruskin wrote:

> *God has lent us the earth for our life; it is a great entail. It belongs as much to those who are to come after us ... as to us; and we have no right by anything we do or neglect, to*

> *involve them in unnecessary penalties, or deprive them of benefits which it was in our power to bequeath.* (Ruskin 1883, 185–6)

The preamble to the UNESCO Convention notes that both cultural and natural heritage 'are increasingly threatened with destruction' and that 'the deterioration or disappearance of any item of the cultural or natural heritage constitutes a harmful impoverishment of the heritage of all the nations of the world'; further, that the Constitution of UNESCO states 'that it will maintain, increase and diffuse knowledge by assuring the conservation and protection of the world's heritage' (UNESCO 1972, 1).

The concept of a valued heritage is normally assumed to go hand in hand with the requirement to protect or conserve it from this impoverishment:

> *The distinguishing characteristics which define heritage and bind it together are:*
>
> *(1) a powerful underlying commitment and unifying set of attitudes to protect and conserve for posterity the objects, sites, flora and fauna, structures and other material evidence of a community's past and present;*
>
> *(2) a shared perception of the intrinsic and cultural value of heritage and the wish to use it to communicate and interpret the past to present and future generations.* (Middleton 1994, 3)

One way to protect the material heritage is to recognize it formally by placing it in a museum or (as is done, for example, in the UK) by listing buildings or scheduling ancient monuments, and (internationally) by designating World Heritage Sites, but recognition must go hand in hand with measures for its material well-being. All heritage structures and objects, from an urban complex to a tiny mollusc, require some form of long-term management in order to prevent or limit damage and guard against loss. However, the need to protect the material heritage for posterity (future generations) should not be allowed to outweigh the need for it to remain accessible, and usable now. Heritage structures may not be protected for ever, the shifting nature of the focus of heritage is seen in the fact that museum objects may be de-accessioned or monuments de-scheduled, though some will pass to another use, for example objects passed to another museum.

Illicit trade in material heritage

In spite of the assumption that cultural heritage should be protected it is particularly vulnerable throughout the world to looting, theft and illicit trade. Much of this looted or stolen material is bought by unwary or unscrupulous private individuals, but some ends up in museum collections. This is not a new phenomenon but the scale of the problem has increased as the trade has grown and become more lucrative. Monuments are defaced to provide sculptural heads or figures, religious buildings are robbed of their paintings, archaeological sites are dug into and plundered, and museums are stripped of their contents. Some countries, such as Mali in West Africa, have lost a horrifyingly large proportion of their cultural heritage, and thus of their cultural identity (Tubb 1995, Vitelli 1996, Brodie *et al* 2000).

Despite attempts to stop this trade through drawing up instruments such as the UNESCO *Convention on the Means of Prohibiting and Preventing the Illicit Import, Export and Transfer of Ownership of Cultural Property* (1970) and the UNIDROIT *Convention on Stolen or Illegally Exported Cultural Objects* (1995), many governments, including, unfortunately, the UK government, have not ratified these conventions, and therefore tacitly support this trade.

Looting has always happened in times of war but this trade is continuous and, much of it, apparently well-organised. Equipment such as metal detectors has increased the efficiency of finding buried metal antiquities, and global tourism, together with new ways of selling stolen material (such as in department stores and through the Internet), have encouraged the process. It is driven by the fact that there is a market for the material (though some of the purchasers are probably ignorant of the damage it is causing), and by the fact that unscrupulous 'restorers' will repair objects, damaged during looting, to make them saleable.

Conservators must be aware of the growth of this trade, since they may be asked to work on stolen or looted material. This is a particular danger when working for dealers or private collectors, but it is also possible when working for a museum which may have inadvertently, or knowingly, acquired illicit material.

Materials conserved

This book concentrates on portable artefacts but it refers to structures on sites, as well as the sites themselves. It also refers to human remains which, because they have been found in archaeological contexts and are still stored or displayed in museums, also concern conservators. These objects and remains may have been found in a number of different contexts, and retained for a number of different reasons, and they may now be housed and used in a range of different locations or circumstances, all of which affect both their potential for study and display, and their conservation requirements.

Many archaeological objects and structures were discovered through excavation of a buried deposit, or opening of a tomb, and this is, of course, true of many or most of the objects from early periods, and objects which form part of early collections. The fact that archaeology is now seen as wide-ranging in time and concept means that materials in different physical contexts, eg underwater remains such as shipwrecks, or standing buildings such as churches, or from more recent periods such as industrial plant and machinery, or from 'other' cultures such as ethnographic collections can also be considered to be archaeological.

Excavated objects

Although archaeological objects and structures now include material which has not been excavated, what distinguishes objects from land-burial and from underwater deposition is that they often exhibit particular problems of deterioration and conservation. Such materials can be divided into those which have come from current excavations so are freshly exposed and may be in the process of investigation and study; those which have been excavated in the recent past and are now stored as part of an archaeological archive composed of material relating to published sites; and those which were excavated or collected in the more distant past and are part of a museum collection. While most of the material in the last two categories is likely to be in store, some of the more 'beautiful', interesting or instructive objects may be on display. Because these artefacts have been excavated at different times and subjected to different conservation treatments and storage situations, their condition and superficial appearance may vary greatly. Freshly excavated material may be at risk of rapid deterioration as it reacts to its changed environment, whereas excavated material which has been stored over a long period may have reached a relatively stable state either naturally, or through conservation intervention.

Excavated structures

Archaeological artefacts also exist in the form of structures exposed on an excavated site. These may be relatively small in size such as pottery kilns or well linings or much larger such as the site itself, composed of a complex of related features such as walls, floors, drains and road-surfaces: two well-known examples are Pompeii in Italy and Machu Picchu in the Andes. Standing structures are also classed as archaeological, ranging from Stonehenge, or the Arch of Titus in Rome, to more modest burial mounds or standing stones.

These structures present special conservation problems because they are exposed rather than protected, though on some sites partial protection may have been provided in the form of a shelter. Difficulties of protection, or the ultimate destruction of the site itself (eg through new building construction) may mean that smaller structures such as kilns or mosaic floors are removed from the site for preservation and display within a museum. The difficulty and expense of preserving excavated wooden structures means that either no attempt is made to do so (the Roman and mediaeval waterfront constructions along the Thames in London) or they are protected, and conserved, inside specially-designed buildings (eg the *Wasa* in Stockholm, Sweden, or the *Mary Rose* in Portsmouth, UK).

Objects in museums

Wide-ranging and often idiosyncratic 'old' collections of archaeological, ethnographic, social and natural history materials are characteristic of many local museums, and similar collections are frequently found in historic houses. Whatever the reason for the initial acquisition by museums these items are now likely to be selected for use in thematic display; whereas in historic houses they are likely to have been acquired as curios and to continue to be displayed in this way. In general, current museum-collecting policies are selective according to agreed objectives such as to cover local history, or to fill gaps in an existing collection. Larger or specialist museums usually make much clearer distinctions between these categories and devote whole galleries to, for example, the ethnography of a specific culture or region, or the history of a local industry. Some collections (of both real objects and replicas) will have been brought together specifically for teaching and research purposes in universities or related training institutions.

Objects in museums are very varied in their conservation requirements, partly because of the different raw materials from which they are made, partly because of their present condition, and partly because of their perceived significance. Their significance affects the ways in which they may be used in the museum (eg for teaching, demonstration, or display) and this in turn affects their conservation.

A current issue for museums (as for other heritage organizations) is the requirement to provide much greater access to the collections. This can range from working with local communities or indigenous groups to develop sympathetic practices for handling, storing and exhibiting their cultural material, to displaying a far larger proportion of collections either through opening stores to visitors or through more extensive temporary exhibitions.

Private collections

The content and composition of private collections is immensely varied, from a few inherited family possessions to deliberately constructed collections of, say, porcelain or postage stamps. Many privately-owned objects, particularly paintings and furniture, have been repaired or restored more than once in the past and their need for conservation may result in part from poor or deteriorating early repairs and in part from present use. Objects in private collections may be particularly vulnerable to the effects of light (which affects organic and coloured materials such as textiles and pictures) since few domestic houses have any means of controlling levels and quality of light; they will also be affected by the wear caused by handling and daily use.

Conservators working in the private sector may work for the antiques trade, or for private collectors. They are normally concerned with returning objects to functional condition (eg repairing furniture or clocks) or to displayable condition (eg porcelain or paintings). This form of conservation is normally described as restoration, and focuses on the continued use of objects as prized domestic furnishings or decoration.

The current context

Thus the context in which conservation takes place is varied and changing. The issues which affect conservation arise from the collapse of the old certainties and the opening up of the study and enjoyment of the past to new and varied interpretation, and to new and different audiences. This opening up has required museums to widen access often by means of making it possible to handle and use objects in innovative ways. The pressures of access mean more exhibitions, and more loans with the consequent need to prepare objects for transport (often internationally); some conservators now spend a considerable amount of their time servicing loans. The status of the material heritage is also shifting and is another issue which affects conservation. While some material is repatriated to its original owners, other material is wrenched from its original context to be illicitly traded, so, at one end of this spectrum conservators are closely involved in preparing material for return to indigenous communities, and at the other they must be ever vigilant not to support theft by working on objects of dubious origin. The changing context requires conservators to review their discipline and practices.

2 Conservation examined

Introduction

The material heritage is important in many different ways to different groups of people. That conservation exists as an activity arises from a desire to prevent the effects of change from obliterating the tangible and intangible assets of the cultural heritage. All the artefacts and structures which make up this heritage are subject not only to the wear and tear of use, and to the depredations of looting and theft, but also to the material changes caused by deterioration. Just like archaeology and museums, conservation has undergone considerable changes in attitudes and practice since the 1970s. If it is to become better integrated into all that the 'heritage industry' represents it now needs to take stock of, and strengthen, its theoretical base. An intention of this book is to discuss both the thinking and the practicalities behind much of what is currently done in conservation in order to provide a basis for further debate and for development of theory.

This chapter looks at the ways in which conservation can secure and enrich the heritage through understanding objects and through preserving them. It sets out some of the issues which affect the practice of conservation, and examines the role of theory, and of standards and principles, in developing conservation of the heritage.

Scope of conservation

Why is conservation needed?

Conservation minimizes the effects of deterioration. Most materials tend to change when exposed to the effects of agents such as water, light, microorganisms: these cause natural weathering processes which break down solid rock to soils, and waste disposal processes which break down and recycle animal and plant litter. These processes do not discriminate between materials in their raw state or after modification, so natural outcrops of stone and carved building stone are subject to similar types of weathering, and leaf-litter on the forest floor and an intricate textile may be attacked by the same fungi and insects. Some processes of manufacture convert the raw materials into a less stable state; for example metals tend to be less stable than the ore they are made from, some papers are much more vulnerable to decay than the raw plant fibres from which they are made (Cronyn 1990).

The processes of decay have long been recognized and some objects and structures have been deliberately made in such a way that the processes are slowed (eg they may have a protective coating of paint, or may be constructed so as to shed water effectively), but however an object or structure is made, some change is inevitable. Indeed, some changes are considered desirable in that they may be considered to enhance the beauty of an object (eg an attractive patina on a metal object), some are undesirable in that they may put the object at physical risk and destroy its meaning, such as crumbling of stone-work, flaking of paint.

The objects and structures which make up the heritage may be particularly prone to change not only because they may be old and have had a long exposure to normal deteriorating factors but because in many cases they have been the subject of interest over a long period, and thus have been much viewed, handled or climbed over, cleaned and repaired.

Ironically, one of the most drastic changes an archaeological object can be exposed to is excavation. Broadly, objects and structures become protected and stable while buried, but excavation induces an abrupt change of environment which can set up a series of damaging changes in the object; therefore the act of discovery is also potentially an act of destruction. Of course all these effects can be counter-balanced by the fact that whenever objects or structures are prized and studied, attempts are made to protect or conserve them.

How does conservation contribute to the study and appreciation of the heritage?

Conservation has an essential role in the process of understanding the past and the 'other' (here meaning the evidence of unfamiliar or 'strange' cultures), and in interpreting them in the present. By clarifying the evidence they contain and by maintaining them in good condition, it helps to make objects available for study and appreciation by specialists and by the public.

Study and elucidation of objects provides information on ancient and pre-industrial technologies and aids understanding of objects of all kinds, and of their possible functions

Figures 2 and 3 Eighteenth-century leather fire bucket before and after restoration (Courtesy of Royal & Sun Alliance Insurance Group plc, and the Leather Conservation Centre).

in past societies. Conservation focuses on ensuring that objects are usable, that they can be viewed, or handled, or set in motion; thus it facilitates access to the cultural heritage. Remedial conservation and restoration processes enable objects or monuments to be interpreted by visitors and so to enrich the visitors' own experiences of the past (Figs 2 and 3). Conservation is also responsible for balancing the need for access and use with the requirement for long-term preservation, so that the cultural heritage survives into the future, for future interpretation, and re-interpretation.

Just as there are many different views of, or interpretations of, the past so there are different possible views of, and uses for, objects. Conservation articulates with a range of different disciplines, and in elucidating meanings of objects conservators often work with different groups including artists, indigenous peoples, art historians, scientists and the public. Together these views should give conservators particular insight into the varied interpretations of objects from the conceptual to the material; from the aesthetic to the historic or the scientific.

Conservation (and restoration) has an important role in bringing the past to life. It has made it possible to raise whole ships from the sea-bed and to preserve and study both ship and contents. It has reinstated the Albert Memorial in London to its nineteenth-century gilded glory. It has made it possible to understand, and even to imagine wearing, Egyptian clothing (eg garments on display at the Petrie Museum, University College London, see Fig 4). It is responsible for treating the Star Spangled Banner in the National Museum of American History, Washington, so that it will survive to amaze future generations, and it has provided specially controlled atmospheric conditions in sealed display cases to preserve the Royal mummies in Cairo Museum.

Figure 4 An ancient linen tunic from Tarkhan, Egypt, 1st Dynasty: after conservation (Copyright: Petrie Museum of Egyptian Archaeology, University College London, UC 28614).

Processes of conservation

Looked at very broadly, the processes of conservation involve two main activities: rectifying the effects of existing or past deterioration, and preventing deterioration. Both these activities depend on assessment and understanding of the tangible and intangible characteristics of the objects themselves. It is the understanding of objects, and the practice of preventive measures which distinguish modern conservation from earlier craft restoration which focused almost exclusively on rectifying (visible) damage on individual objects.

The analogy is often made with the practice of medicine in order to explain what conservation is and how it works. This analogy is useful when considering the treatment of existing and past deterioration. Conservators working on sites and in museums deal with the remedial treatment of individual objects, just as a medical doctor advises and treats individual patients. The general processes are very similar. After initial examination, both preventive and remedial measures may be used, specialist advice may be sought, and treatments may be reviewed and changed, but the process always includes assessment, diagnosis and documentation. The treatment recommended by the medical doctor is usually intended to enable the patient to live a normal life, ie remain active and involved. Equally conservators choose treatments which allow the object to be used in a number of possible ways, eg in a museum, or other heritage context, or by a private owner. Enabling an object to continue to function effectively may involve some restoration, eg replacement of missing or damaged features. This, too, is analogous to modern medicine, as in the development of organ transplants and plastic surgery.

Where the two disciplines could be said to diverge is in the understanding of the object, and the degree to which preventive measures are applied. A common complaint about conventional medicine is that it fails to consider and treat the 'whole person' (both physical symptoms and emotional needs) or to take his/her own views into account. In conservation, increasing emphasis is put on establishing both tangible and intangible significance of objects and collections (both how they are made and what they may mean), as well as on consulting interest groups; information derived from the object and from consultation is then taken into account when making conservation decisions.

A large proportion of conservation work involves applying preventive measures, attempting to slow the rates of deterioration or to stop the onset of deterioration, and there is increasing emphasis on the management and care of whole collections. Indeed it could be said that now all remedial conservation takes place within a framework of preventive care. Preventive medicine and health education are similar to preventive conservation in that they focus on the health of groups or populations, and on modifying or controlling external factors such as pollution, sunlight, smoking, diet and spread of epidemics. However, unlike the situation in conservation, preventive medicine is still given less emphasis in conventional medicine (in the UK at least) than remedial treatment, perhaps because it is only successful in the care of small children where the parent makes decisions (adults can, and do, decide to ignore health guidance). Thus the conservator's role as an advocate for the preventive care of objects could be compared to the role of a parent in supervising the health of children as yet unable to speak for themselves, but it is worth adding that it is as dangerous for conservators to become over-protective of objects as it is for parents to become over-protective of their children.

Conservation deconstructed

Conservation is concerned with safeguarding objects or structures now and into the future. It involves assessing the state and meaning of the object and applying procedures which will remedy existing damage and/or prevent future change.

While, in practice, conservation is concerned with minimizing the most serious and damaging changes, and maintaining what remains of the object or structure, what it is actually aiming to do is to safeguard the characteristics of the objects or structures which make them informative, awe-inspiring or enjoyable, that is, their significance. In some cases, particularly with highly degraded excavated artefacts, the significance may become apparent only during the procedures of examination and conservation (Fig 5). Furthermore it is known that there are some aspects of objects which are not yet understood (Keene 1980, Ward 1986, Cronyn 1990, Pye 1992). Thus the shape and function of a corroded bronze object, as well as the decoration on its surface, may not become apparent until it has been investigated and cleaned. Thus, too, the possibilities of discovering the types of food-stuffs once contained in ceramic jars or amphorae are increasing, but there is probably more to discover.

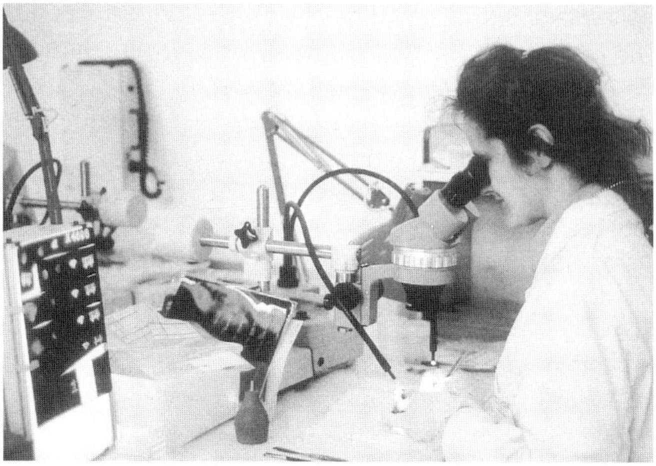

Figure 5 An archaeological conservator examining an object under a microscope during the process of investigation (Courtesy of the Museum of London).

Obviously it is normally conservators who do the conservation work but it is clear from this analysis that others will be making an important contribution in defining the significance of the object – this aspect has to be a team effort. The external factors which conservators deal with are deterioration factors and processes, some of which can be prevented by controlling the environment around the object. In preventive conservation, the work of the conservator will again be shared with others (curators, collections managers, registrars, warding staff, technicians, owners) who have a responsibility for maintaining a conservation environment.

The people who benefit from the conservation are those who 'use' objects, that is both the academic community and the public of all kinds (and owners of objects or buildings which are in private hands). The people who control the process are the owners (in the public domain these are the museum and heritage managers), and, in some cases, the originators, who may include indigenous groups, and living artists.

Terminology

Conservation is difficult to explain since there is a tendency to use 'preservation' interchangeably with 'conservation'. Usage varies in different contexts and different languages, although 'conservation' is becoming a more widely recognized term. Two main confusions involve firstly, the use of preservation to mean what is now more generally called conservation, hence 'Historic Preservation' in the USA (Fitch 1982), and secondly the interchangeable use of 'conservation' and 'restoration'. In many European languages there was no distinct term for conservation as it is defined here and the equivalent of 'restoration' and 'restorer' have been used to cover both restoration and conservation (in French 'conservateur' means curator). Indeed, it is common to refer to restorers of paintings but conservators of archaeological materials, even though both may be fulfilling similar functions. This confusion has led the Conservation Committee of ICOM to coin the term 'Conservator-restorer' (ICOM-CC 1984).

Conservation or preservation?

The meaning of conservation, as used in this book, becomes clearer if it is contrasted with the broad meaning of preservation. The Oxford English Dictionary definition of conservation is 'preservation from destructive influences, natural decay or waste; preservation in being, health, perfection, etc', and of preservation is 'the action of preserving or keeping from injury or destruction'. Although these definitions are apparently similar there is a crucial difference between conservation and preservation which relates to the degree of protection, and to use. Preservation tends to mean protection from all harmful or damaging factors and one thinks of things sealed up or isolated, and certainly not in daily use (hence 'preserved in amber'). However, conservation means protection only from unnecessary harm or waste. A supply of food which is preserved is, by definition, not in use (until the seal of the jar is broken, or the salt is soaked out). A supply which is conserved is protected but in use; in other words it is used carefully and gradually (and may eventually be used up). Conservation therefore implies responsible and cautious use, whereas preservation implies total protection. Objects are not expected to last for ever. 'Objects, like ourselves, have a finite life-span, although their lives are frequently much longer than ours are' (Pearce 1992, 16).

Preserving an object or structure is like protecting a very sick person in an isolation chamber: the object may be 'alive' but this is not normal life, and interested people cannot interact with it. There are some special cases where uniquely valuable objects are treated in roughly this way, eg the American Declaration of Independence, although on display, is sealed in an atmosphere of helium (Thomson 1986) and some early plastic objects have to be kept in refrigerators or deep-freezes to prevent deterioration (Reilly 1991, Fenn 1995).

Preservation of museum collections can also be seen as isolating them, since the act of placing objects in a museum represents a deliberate decision to take them out of their normal use and to transform their status to museum object, ie to something protected from the hazards of over-use, modification, accidental breakage, and discarding when worn out, or out of fashion. The museum is, in essence, a sealed environment but one within which a number of specifically museum activities takes place, including the conservation of objects for use (or the controlled demonstration of function, eg of early machinery). Here use may not mean the original intended use, but a museum use, ie being able to see, study and enjoy the object.

It is also useful to contrast conservation with the preservation of archaeological objects in the ground. During burial or entombment these objects are removed from normal use and everyday deterioration and are isolated in sealed and more or less unchanging environments until they are excavated. In some cases it is not just the safety of isolation that has preserved the objects and structures but truly preservative conditions such as desiccation, or very low temperature. Examples of preservation in this latter sense can be spectacular, such as the goods entombed with Tutankhamen, or the Copper Age 'Ice-man' who was sealed under the ice in the European Alps (Spindler *et al* 1996). Once objects are removed from these conditions they need conservation attention in order to perpetuate their preservation, either through maintaining similar conditions in a new context, or by treating them to enable them to survive in normal museum conditions.

Conservation and the object

Objects have different significance for different people. The same object may be appreciated for its colour or surface quality by one person, for its demonstration of early technological ingenuity by another, or for its intangible religious attributes by a third. The object is sometimes compared to a written document which has a message, or a number of messages, to impart. Each user is 'reading' the object and deriving information or pleasure from it. Choosing suitable approaches to conservation involves making sure that the treatment chosen safeguards all the information in an object, and does not conserve one aspect of the object while adversely affecting another and thus distorting the significance.

The significance of an object has both tangible and intangible aspects. The tangible aspects concern everything relating to the materials and manufacture of the object, signs of use and age, and its present condition. These are qualities which a conservator can hope to elucidate by examining the object, and by applying various tests (and by seeking the advice of conservation or materials scientists). However, there is always a limit to the amount of information any one person can derive from an object at any one time. The intangible qualities are difficult to assess and involve a range of views about what is important about an object. These qualities may be partly evident through examination (eg the aesthetic qualities of a painting) but more often they are understood only by those who have a direct cultural link to the object (as in the case of a scroll used in religious ceremony, or a garment worn during initiation rites).

Building up a picture of the object is therefore an activity shared between conservator, curator, materials scientist, and interest groups such as original owners or makers. The material state of any object requires some investigation, but investigation becomes particularly important when assessing excavated objects which may be impossible to interpret without detailed examination, and removal of superficial accretions. The terms 'investigative conservation' and 'investigative cleaning' are sometimes applied to this assessment of archaeological objects. The results of investigation aid archaeological research, influence the way an object may be used, and so affect choice of conservation procedures:

> *conservation research is no different from, nor less important than research in any other discipline (archaeology, science, cultural history etc) by which an object can be understood. It is the sum of all these fields of enquiry which provides the understandings of an object's materials, rate of deterioration, context, original use, date, position in science, art and history, which in turn forms the basis for any museum decision affecting it. (Hebditch 1998, 10)*

However, it is not possible to establish all the information within an object, or all the possible views of its significance. The means may not yet exist for discovering some of the information but conservators have to be aware of the potential for all objects to yield further information, and meanings, when new research questions are asked and when new methods of investigation are developed. Conservation processes must aim to safeguard both potential as well as apparent significance.

Conservation and use

Conservation ensures that the object is accessible and usable. Use and accessibility means accepting that the object may be handled or exposed to light, or (in the case of buildings and structures) exposed to the weather, even though it is known that these actions are potentially damaging. Conservation aims for reasonably stable health rather than no change at all. It involves assessing the possible risks inherent in use and limiting them where necessary, rather than eliminating all risk (except, possibly, for unique and highly valued objects).

While many objects are no longer used as originally intended (costume no longer being worn, swords not used for fighting, pots not used for cooking), in some cases use is not much different from the original use but museums or heritage centres provide controlled access to the objects for study and enjoyment (paintings and tapestries may still furnish a room, machinery may still be put in motion). Some protected or conserved buildings (eg in the ownership of the National Trust in the UK) are still lived in. Conserving for use inevitably involves compromise, and acceptance of some change within the object.

Conservation and restoration

The definition of conservation given above indicates that the focus is on minimizing existing, or future, effects of change. This can be contrasted with the activities which are normally described as restoration, which focus on revealing and clarifying or making more understandable the attributes (usually visible attributes) which contribute to the significance of the object. Restoration is intended to aid interpretation of objects. In this sense, restoration is distinct from conservation because it does not focus on controlling deterioration, so does not necessarily contribute to long-term preservation. However, many conservation activities contribute to restoration and some conservators would not make as clear a distinction between the two activities (Ward 1986, Oddy 1992a, Oddy 1992b).

Restoration is familiar in the form of treatments applied to objects such as paintings or tapestries where the intention is to make the image understandable, but it can also apply to working objects such as clocks or industrial machinery where the intention is to restore the function by putting the object back into working order. Restoration is strongly linked to the way in which an object may be used in a museum (or in a wider heritage context).

The extent and purpose of restoration depends on the significance and use of the object. Restoration may be done for mainly aesthetic reasons (eg completion of the image through restoration of missing elements in paintings) or for mainly didactic reasons (eg reconstruction of an archaeological pot to show its original shape and function), but it is almost always linked to the intention to look at or display the object. Although the objectives of conservation and restoration are different, the processes lie on a continuum (see Chapter 7). The fact that a conservator of paintings or other art objects is often referred to as a restorer does not mean that s/he does not conserve, but indicates the importance attached to restoring the artist's original intentions, and of being able to appreciate the aesthetic whole, both considered crucial to the understanding of art works. Thus conservators of art works are normally also skilled restorers. It could be said that there is a professional continuum from conservator to restorer, with, for example, archaeological conservators at one end (who are seldom engaged in restoration) and paintings restorers at the other end (for whom restoration is part and parcel of the conservation process).

Remedial or preventive conservation?

The earliest types of conservation were applied to objects in order to remedy (cure) existing damage, to repair broken parts and to replace missing features. It was not until the nineteenth century that scientists began systematically to study deterioration and its causes, and not until the twentieth century that scientists working in the conservation field began to propose measures to prevent deterioration.

Remedial treatments almost always involve direct intervention with an object (eg removing incrustations, repair or reconstruction of broken fragments) and so they change the object

(Oddy 1992a). Many preventive procedures focus on controlling the environment around objects (eg controlling exposure to light or to humidity) in order to minimize deterioration, and so do not change the object itself (Knell 1994, Cassar 1995).

The term 'preventive conservation' is usually taken to cover only measures which control the environment, and activities such as handling or transport of objects. However, the point of conservation, as opposed to the old craft-restoration, is that it anticipates further deterioration and involves procedures which safeguard objects into the future. Remedial treatment should certainly not exacerbate deterioration, and most remedial treatment has the intention of stabilizing the object, ie putting it into a condition which will allow it to survive in the same state as long as it is in a controlled environment. The term stabilization is often used to describe the halting of deterioration using either preventive or remedial measures.

There is another potential confusion in the term 'preventive conservation' in that some treatments that involve interaction with an object (such as removing potentially damaging salts) have the primary objective of preventing future deterioration, rather than remedying any existing degradation. Thus there are two main activities within the framework of preventive conservation, one of which involves controlling the environment and activities surrounding the objects, and the other of which involves direct intervention in the object itself. These should perhaps be distinguished by calling them 'preventive care' and 'preventive treatment'.

From preservation to disintegration

All conservators accept that not only can some objects be conserved more successfully than others, but that some will inevitably deteriorate even in the most carefully controlled conditions. It is accepted that some uses will hasten deterioration (keeping clocks going, using objects in handling collections). In some cases conservators have to face the fact that objects were never intended to last (newspapers, bus tickets) or were actually intended to decay (objects which are discarded or allowed to decay as part of a ceremony or ritual (McLeod 1995)). Without deterioration the world would be stifled by vast numbers of objects and structures of all kinds and all ages, and study and enjoyment would be compromised, if not impossible. Deterioration, and for many objects death, are inevitable (even desirable), though they may take place over an extended time-scale.

Thus the possibilities of safeguarding heritage objects and structures lie along a continuum, starting with preservation, moving through conservation and different degrees of change, and finishing with decay and disintegration. At one end some types of object are preserved (eg anatomical collections sealed in alcohol or formaldehyde, which can be examined only through the glass, but not exposed or touched) and at the other end some objects are not considered appropriate for conservation so may be allowed to deteriorate (Gathercole 1995).

Conservation and materials science, or conservation science

The work of the conservator and the materials scientist often overlap: when working on complex objects the conservator may raise questions for the materials scientist, and may need further information from the scientist before conservation work can be taken further. Some objects are difficult to interpret when first assessed and more detailed information is needed on structure and composition before it is possible to decide on the material significance, and thus on appropriate conservation procedures. Materials scientists specialize in the study of materials and their behaviour, eg reactions of clay bodies during firing in ceramic manufacture, or the mechanisms of corrosion of metals; some may be involved in the development of new materials such as synthetic polymers. They may be able not only to identify materials but also to interpret structural variation in terms of likely technologies or processes of deterioration. To do this they often use instrumental examination or analytical techniques, such as scanning electron microscopy or X-ray diffraction, in order to provide a level of information which is not available to conservators working with normal conservation equipment such as binocular microscopes.

Some scientists, usually known as conservation scientists, specialize in supporting, researching and developing the processes of conservation. They may study ancient materials, but they may also be involved in evaluating modern materials such as adhesives or display materials which are relevant to conservation. Conservation scientists test existing conservation procedures, assess techniques used in other fields for their applicability to conservation, and, where possible, develop new procedures. Unfortunately not all their work is directly relevant to the day-to-day problems of practical conservation, thus conservators often do their own testing and evaluation of materials and processes. There are comparatively few conservation scientists since posts exist only in a few large institutions such as universities or national museums. This means not only that many conservators do not have access to the advice and support of a conservation scientist, but not enough scientific research and development work is being undertaken in conservation because there are not enough conservation scientists. Few conservators have the time and equipment to pursue in-depth research.

Who is responsible for conservation?

There are three main groups of people who are involved with, or responsible for, conservation. The first is the public, including owners and makers (or their successors). Many members of the public have been instrumental in 'saving' buildings or monuments which might otherwise have been demolished, and in setting up local or national bodies which are responsible for conservation in the broad sense. Owners of objects may commission conservation, and the desire to own and display, or use, old things creates a demand for conservation. Makers such as contemporary artists, or their successors (eg indigenous groups wishing to reclaim their cultural heritage) may wish to stipulate whether, and how, conservation should be undertaken. Involvement of the public stimulates conservation and is essential if there is to be support for, and 'ownership' of, heritage conservation.

The second group consists of professionals who work with conservators. Many heritage professionals have a broad responsibility for conservation, eg field archaeologists, who are responsible for seeing that conservation is built into any archaeological project from initial assessment right through to strategies for preservation *in situ* (where viable). Curators and collections managers are responsible for ensuring good preventive care of museum collections. These responsibilities are clearly stated in codes of practice for archaeologists and museum professionals (eg ICOM 1996, IFA 1999). However, archaeologists themselves are involved in conservation of objects only when first aid is needed for objects during excavation. Not all archaeologists do consult, or involve, conservators in the course of their excavations (Stanley-Price 1989) though in some situations, such as the excavation of waterlogged sites, on-site conservation will be necessary to prevent very rapid loss of information through structural collapse of wood or other materials as they dry. When an archive of excavated materials passes to the museum it is the curator and museum conservator who are responsible, but museums (certainly in the UK) will require that the material is stable before they receive it.

The third group consists of the conservators themselves. Conservators have varied roles: most, of course, carry out practical conservation work, but increasingly the process of conservation involves wider management of resources and may include environmental or collections management, policy development within public organizations (eg in Resource: the UK Council for Museums Archives and Libraries), or teaching conservation within universities. Most conservators have either direct or indirect responsibility for the care and conservation of objects, and most contribute to policy development or management plans even if they are not themselves responsible for them. Conservators working on objects or collections may be advised by conservation scientists, and may be assisted by conservation assistants, volunteers or technicians working under their supervision. Although the practice in building conservation is slightly different (conservation architects plan and manage remedial work but the work itself is undertaken by specialist builders), normally it is only conservators who undertake work which involves direct intervention on the object itself.

Ethical and theoretical basis of conservation

Conservation, just like medicine, has the power to damage, distort, or destroy the objects it is intended to care for, thus, again like medicine, it is practised within a framework of ethical standards and principles. Principles and codes of practice represent attempts to set standards and to guide or regulate practice. In conservation this is done to protect cultural heritage from irresponsible or damaging practice, and to protect the interests of the public, owners and so on. The existence of principles governing conservation is another feature which distinguishes modern conservation from earlier craft-restoration.

There is an ever-widening range of international charters, of documents setting out standards or principles, and of codes of practice, relating to the understanding and preservation of the heritage. Many of these have a direct or indirect impact on conservation. Examples include the UK Institute for Conservation *Guidance for Archaeological Conservation Practice* (UKIC 1990), the American Institute for Conservation *Code of Ethics and Guidelines for Practice* (AIC 1996), the *Venice Charter* (ICOMOS 1966) and the *Burra Charter* (Australia ICOMOS 1999). Those applying to sites and monuments have been recently reviewed by Bell and those relating to conservation in museums by Sease (Bell 1997, Sease 1998). It is also interesting to note that in the field of archival conservation in the UK core principles have been set out in the form of three British Standards (Pickford *et al* 1997).

Principles of conservation

There are several principles which are common to different fields of conservation, and which have been directly or indirectly enshrined in various codes. The first is to ensure that any work undertaken should not affect the significance of the object, either the material structure or the intangible meaning. The second is that the minimum amount of interventive work should be undertaken on an object which will secure a satisfactory result. This is often linked to the intention that, where possible, new materials introduced should be removable (the old principle of reversibility). A third is that when any restoration is undertaken it should be readily detectable, even if not immediately apparent on superficial examination. Conservators are expected to work only within their own expertise (ie an archaeological conservator should not attempt to restore an easel painting, and a paintings restorer should not attempt to investigate and conserve an excavated metal artefact), and all work should be fully and clearly documented. Codes and charters vary in their emphasis and level of detail, and some, of course, set out variations on these general principles to suit particular contexts (eg conservation of monuments or of objects). Conservation in its broad sense is also covered by a number of international charters which deal with issues such as cultural tourism and the effects of war.

The nature of principles

The Oxford English Dictionary provides more than one definition of 'principle': here principle is taken to mean 'a general law or rule as a guide to action' rather than 'fundamental truth'. Principles in this sense provide a useful indicator of the thinking at the time that they were set down, and, as Sease points out, examining changes and revisions to conservation principles provides a picture of the development of the discipline (Sease 1998). However principles can lag behind current thought and may even stifle new ideas.

How useful are principles in guiding practice? Principles are frequently seen as inflexible and restrictive, and they may be seen as prejudices. Lasko put it very nicely when introducing a discussion of the principles of restoration:

> *I am well aware that principles are awkward things, liable to arouse opposition by their mere existence. But I intend to make the attempt, and to stay firmly within the context of history, rather than enjoy the luxury of placing before you my own set of private prejudices – although that is what principles are before they are accepted by a group, be it amenity society or nation. (Lasko 1992, 143)*

Another difficulty with principles is that they may be seen as too idealistic, not rooted in the real world. It may be relatively easy to discuss and develop a set of principles but trying to apply them in practice may prove very difficult. In other words, the ideals defined in a set of principles may be difficult to adhere to in practice. An example of this is the concept that only a single standard of conservation should be applied, regardless of the importance of the object. This was set down as a principle in early versions of the American Institute for Conservation's code of ethics but it posed considerable problems for conservators having to deal with large quantities of material, eg in the fields of paper and archaeological conservation (Sease 1998). It is now widely accepted that some selection has to be made because there are simply not the resources to conserve everything to exactly the same (high) standard.

Principles are not static. Principles change to reflect development of knowledge and practice within the subject, the effects of external events and the effects of fashion. Many of the conservation principles which emerged during the nineteenth century resulted from public outrage at restoration of ancient buildings, many that emerged in the twentieth century were a response to the direct or indirect effects of war. Principles reflect thinking which may have been developing over some time, thus they may exist in different forms. Some may have reached the stage of being formally accepted and codified, some are beginning to take shape and being tested through discussion and publication, and some are being re-evaluated.

It is a weakness of principles that they may fossilize thought, and evolution of ideas. This was the case with the principle of reversibility which was eventually dropped from several codes because it was acknowledged that it is impossible to achieve, but it had been such an important tenet that it was some time before its weaknesses were fully discussed. Only very recently was a conference organized on this topic, during which speakers brought together current thinking in the field and considered whether the principle should be retained, modified or discarded. This will almost certainly prompt further discussion, and further refinement of existing codes of ethics and practice (Oddy & Carroll 1999). Conservation codes have also been slow to acknowledge formally the importance of intangible significance and its effect on conservation of objects and monuments (McCredie 1991, Clavir 1994). This has led, particularly, to considerable difficulties over the designation of World Heritage Sites, since intangible significance was not, until recently, considered to be an allowable criterion for selection (Larsen 1995). At some time in the future the current predominance of preventive conservation may need to be re-evaluated since it is possible that it is stifling research into better methods for remedial conservation, and even for restoration (both of which remain necessary).

Theory and principles in conservation

A discipline develops through stimulation of thought and ideas, and through debate in specialist literature and meetings, but conservation of archaeological and other objects (eg social history) has not yet developed its theoretical base to the same extent as the conservation of easel paintings and monuments. Archaeological conservation needs to discuss and develop its theoretical base for the same reasons that archaeology needs to 'make an impact in current society and in competition with other disciplines. It is through theory, which systematizes and forms a body of knowledge according to specified principles, that the discipline takes its form' (Hodder 1992, 6). Hodder further argued of archaeology (and the same can be said of conservation as an aspect of archaeology):

> *if archaeology as a discipline is to act effectively in relation to a quickly diminishing heritage, and if it is to achieve a wider public participation in the past, it must claim a coherence. While it might be argued that such coherence can be given by common methods and techniques, the decision about which methods are allowable ... can only be based on theoretical considerations. (ibid)*

Much of the published work which has developed a theoretical base for conservation has come from the fields of monument conservation and of easel paintings restoration, as is clearly indicated in the papers collected in the volume entitled *Historical and Philosophical Issues in the Conservation of Cultural Heritage* (Stanley-Price *et al*, 1996). In comparison, the theoretical discussion of conservation of archaeological objects is little developed.

Conservation has evolved from the activities of craftsmen and restorers who were guided by their skills, and knowledge of the working properties of materials, rather than by theories or principles. By the middle of the twentieth century conservators were beginning to formulate guidance for practice and professional conduct. One of the first sets of conservation standards to be published was the Murray Pease Report and in introducing it Dr A van Schendel (then President of the International Institute for the Conservation of Historic and Artistic Works) was quoted as saying 'one of the important distinguishing features between the craft and the professional phases is that in the latter there is an accepted corpus of principles and methods' (Murray Pease 1964, 116). Conservation is now moving into this professional phase which is indeed characterized by an interest in clarifying issues and establishing standards and principles.

The existence of principles should challenge conservators to think about the implications of what they are doing, and require them to examine their actions critically. In fact, examination of any set of principles demonstrates that their application requires individual judgement and interpretation: they are not comprehensive rules but guidance on what factors should be given priority when making decisions. Lasko (1992) indicated (in the passage quoted earlier) that principles had normally been discussed and refined by a group, so they result from the comparison, selection and distillation of a variety of views. Referring to a set of conservation principles can thus be compared to referring to the ideas and advice of professional colleagues. Potential decisions can be measured against the principles; the principles will not necessarily provide answers but may help to indicate which answers would generally be considered (by the community) to be the most nearly right.

However, it is a weakness for conservation as whole that there are now so many different charters, codes, documents and recommendations, each of which addresses a different aspect of conservation, or a different audience. Most of the international charters refer to the conservation of monuments and sites, and seem intended to reach government agencies, politicians and administrators, as much as members of the profession. By contrast, codes referring to the conservation of moveable objects are largely 'internal' documents in that they are prepared by professional bodies for their own use. Although many of the principles are common to most of the codes and charters, this multiplicity of documents makes it appear that there is no coherent view across the different fields of conservation.

Current issues in conservation

Discussion of theory and principles is important if conservation is to make an effective impact within the heritage field. Conservation needs to develop a coherent view by agreeing the common ground between the different fields and specialisms of conservation rather than perpetuating the differences. It is now considered inappropriate to separate the conservation of a landscape from the conservation of the monuments within it as the two are interdependent for their meaning and impact. It is equally inappropriate to distinguish two separate fields of conservation relating to moveable and to immoveable objects when many immoveable objects, eg buildings, contain collections of moveable objects (indeed may have been built to house them), and this thinking is now embodied in the *New Orleans Charter* (APT & AIC 1991). However, as the conservation field has grown, and techniques of practice have become more complex, the preoccupations of increasing specialism (easel paintings, wall paintings, polychrome sculpture, textiles, costume, upholstery, and so on) have tended to encourage conservators to communicate only within specialist groups, and there is a danger that the common ground is no longer recognized. Development of different specialisms should not take place at the expense of collaboration and cross-fertilization.

Conservation needs to develop both coherence and self-confidence, but not at the expense of new ideas and variety of views.

> *The difficulty is that any such coherence, while it may have the advantage of empowering the discipline as a whole to play an effective role in society, threatens to stifle diversity,*

> *critique and change. Such coherence and self-confidence may disempower alternative and subordinate voices.* (Hodder 1992, 6)

Conservation can be seen to have some disadvantages. Taken in the broadest sense, and applied to the safeguarding of all excavated and other artefacts, it has supported the retention of almost all material from recent excavations in the UK, and thus has helped to create enormous storage and maintenance problems. At the level of stored material conservation is not discriminating and the same basic care can be, and often is, given to everything. 'Blanket' conservation may be used as an excuse for not making decisions about significance, selection, finding new uses, discarding or de-accessioning (although this is now changing) (Merriman & Swain 1999). The protective measures applied to objects can be seen as making objects less rather than more accessible because, for example, conservators may wish to limit handling of objects, or exposure to light. Collections packed away in well-organized stores are not readily accessible to the public. Another effect of the modern conservation ethic, with its emphasis on preventive measures, is the apparent down-grading of remedial skills and stifling of restoration and even of creativity (in the sense of making new objects). Conserving what is there, however crumbled and unrecognizable, may be used as an excuse to rule out skilled restoration, or the creative reinterpretation involved in carving a new statue for a niche in a cathedral.

Conservation today

Conservation has an important contribution to make to the care and use of the heritage, but its weaknesses need to be discussed and addressed. It has developed from a traditional craft approach; practitioners have been preoccupied with developing practice rather than discussing theoretical issues, and in strengthening the discipline and promoting the development of a conservation profession rather than collaborating with colleagues and the public.

Although this situation is changing, cultural heritage conservation has adopted a protective stance rather than interacting with the public, so has made little impact on the public consciousness compared, say, to the conservation of the natural heritage. These are all issues which need wider debate; some are discussed in later chapters of this book.

3 The history of conservation

Introduction

Over the past few years there has been an increasing interest in the history and development of conservation. This seems to arise out of conservators' desire to take stock of their progress, to reflect on where conservation has come from and where it should be going in the future. Within the field of conservation, practitioners do not necessarily agree on what its key functions should be (or even whether it is grounded in art or science); in fact, conservation draws on many different traditions and makes use of skills and techniques developed in a range of different fields. Reflecting on these different traditions shows why it has taken some time to begin to develop an agreed purpose, or clear identity.

Another important aspect of looking back at what has been done in the past is to improve understanding by today's conservators of treatments that have been applied in the past and the effects they have had on objects. Knowing more about early treatments and early materials will improve future practice; it has a bearing on factors such as choice of effective retreatment, health and safety, and ethical issues such as whether to retain or remove earlier repairs.

It is possible to identify two main roots of conservation, the first involving adapting and developing existing knowledge and skills deriving, for example, from craft technologies, and the second involving the development of new approaches and techniques, often in response to new problems. For example, the conservation of textiles has, in part, grown out of the domestic techniques of mending and support for damaged textiles, and the conservation of furniture has made use of cabinet-making techniques and materials. However, corroded and encrusted excavated archaeological objects presented new problems for which there were no established techniques, thus new approaches and materials were adopted.

Roots of conservation

Early understanding of deterioration

'Lay not up for yourselves treasures upon earth where moth and rust doth corrupt' (Matthew, vi, 19). This first-century AD biblical advice encapsulates the transitoriness of material things and accords with evidence of early attempts to preserve. From earliest times decay must have been a readily observable phenomenon, as were some of the ways in which it could be prevented. It must have been obvious that materials which were kept dry or cold survived better than if they were damp or warm. The earliest evidence of attempts to intervene and prevent decay relate to the protection of foodstuffs, and the ritual preservation of (human) bodies. Deterioration of foodstuffs presented a threat when a regular supply of food was impossible to guarantee. There is evidence for food preservation such as removal of water by drying or salting (Forbes 1955 onwards) and for protective food storage, eg structures for grain storage which were apparently intended both to keep the grain dry and to isolate from pests (Pliny the Elder XVIII). These approaches which attempted to control the environment surrounding the foodstuff are similar in method and effect to what is now called preventive conservation (preventing or slowing the rate of decay by changing the environment, but not necessarily making a direct change to the object itself).

Survival of the human (or animal) body into the next life was another preoccupation of many early cultures. Mummification provides an example of successful early preservation techniques, such as the Egyptian techniques involving drying (using natron – sodium sesquicarbonate – as a desiccant), disinfection with strong smelling substances such as myrrh and spices, and a protective coating of resin or bitumen (Cockburn 1980, Peck 1980, Keck 1996, Spindler *et al* 1996). Mummified bodies from other areas and periods show similar preservative factors, whether intentional or accidental. Mummification in the Torres Straits and in Papua New Guinea, for example, involved exposing the body to the air to dry it, sometimes coupled with smoking, which both dries and has a deterrent effect on microorganisms (Pretty & Calder 1980). In the Canary Islands bodies were similarly treated with the addition of pine-needles, presumably for their aromatic qualities, and fats, probably to provide a water-resistant coating (Ascenzi *et al* 1980). In most cases, successful preservation was aided by a dry climate (either hot or extreme cold).

These techniques show that there has long been a good empirical understanding that removing water slows or stops deterioration, that a water-resistant coating can offer further protection and that some materials (such as aromatics) are a deterrent to microorganisms and insects. These have parallels in modern techniques of conservation.

It is interesting to note that attempts to preserve animal skins and furs for utilitarian purposes involved similar methods and materials, and established some early forms of leather manufacture (involving chemical stabilization of the collagen in the skin). Drying using air or salt was commonly used as a first stage to prevent the onset of putrefaction, and application of alum (potassium aluminium sulphate) and fats together resulted in a form of soft white leather which was widely used in ancient Egypt (Forbes 1955 onwards). Smoking, and treatment with fats and oils were also discovered to be effective methods of leather manufacture and many Native American objects were made in this way (Thomson 1981). Vegetable tanning, the most widely used form of preindustrial leather manufacture, however, may have been discovered in an attempt to dye the skin rather than to preserve it.

Traditions of domestic repair and maintenance

The repair and maintenance of household or other mundane objects is another of the roots of conservation. Before industrial mass-production the cost of making things, both in time and materials, was comparatively high, so many objects and structures were repaired, remade and finally recycled to get the most out of the raw materials (eg textiles, wood) or individual components (eg metal fittings). Textiles provide good examples of both remaking and recycling. Textiles frequently survive on archaeological sites only in the form of worn scraps, presumably resulting from discarded garments (or other textile artefacts) being reused as household rags. On the other hand, garments made of valued textiles such as patterned silks were apparently often remade to follow the latest fashion, the later garment showing the cutting and seam-lines of the earlier one (Doré 1978, Brooks *et al* 1994).

Most early domestic repairs were probably undertaken by the users of the objects rather than by specialist craftspeople and were apparently intended to keep objects in use as long as possible, so were not necessarily intended to ensure long-term survival. However, the methods show a good empirical knowledge of the behaviour of materials, and many of the general approaches continue in use or have been adopted and adapted for conservation purposes. Simple mending of many types of object was a task anyone might attempt; the exceptions would have been the repair of materials which were difficult to manipulate without specialist skills (eg soldering or riveting) or so highly valued that they were considered to require specialist attention. Textiles again provide good examples of this type of domestic repair. Many excavated textiles show signs of mending such as couching, patching or darning and there is particularly good evidence of this from Egypt (Janssen 1986). This tradition of mending or replacement using similar materials (ie threads and textile) has continued unbroken as domestic practice and has been adapted by textile conservators (Landi 1985).

Some materials, such as broken ceramics and glass, could not be repaired using the materials of manufacture (though missing pieces could be replaced with material taken from another object). Instead an adhesive was needed to hold together the broken fragments. Adhesive materials such as plant resins or bitumen were exploited from Neolithic or earlier times, and were used, for example, in the construction of artefacts such as composite flint tools in which the flint 'teeth' were held in place with the adhesive (eg Bar-Yosef 1985). The technique of repair using adhesives was also an early development and there are examples of ceramics dating from as early as 7,000 BC which have been repaired with bitumen (Williams 1988). Domestic repair using adhesives was commonplace in later centuries and early household reference books frequently include recipes for cements amongst the advice on health, wine-making etc (eg Holland 1825). Adhesives have remained one of the main methods for reconstruction and restoration of ceramics, glass and stone objects. An alternative joining method using metal rivets or dowels is found frequently on Greek and Roman ceramics (Koob 1998), a tradition which continued well into the twentieth century, for instance in the UK, with itinerant workmen going from house to house mending broken china. Metal armatures were also used to support restorations. Metals are not normally used by today's conservators but early repairs or restorations using these methods might now be considered worth retaining as examples of early techniques.

Another aspect of domestic maintenance was keeping objects clean of matter which might stain or cause other forms of deterioration. Textiles were washed and, until recently, a modified form of washing was very heavily relied on as a conservation technique, and is still used on a limited scale. Removal of surface corrosion from metal objects through polishing with a mild abrasive is also a long-standing tradition. Classical bronzes are thought to have been cleaned to keep them free of corrosion; this would probably have been true of domestic utensils and seems also to have been true of statuary (Craddock & Giumlia-Mair 1993, Hughes 1993). Ancient jewellery was bright and shiny, though in this case surfaces were probably burnished, rather than polished with an abrasive (Ogden 1993); however, it is thought possible that ancient Greek silver may have been allowed to acquire a stable smooth black tarnish layer (Vickers 1995). Silver was polished in more recent times, as can be seen from surface wear, and cleaning silver is customary today both in a domestic and in a museum context. (Conservators use a chemical method for removing corrosion layers because abrasive polishing inevitably removes some of the metal itself.)

Domestic housekeeping guides produced during the nineteenth century in the UK brought together examples of advice and recipes, many of them probably already old, for caring for and repairing a variety of objects. Investigation of these guides has shown that cleaning paintings was apparently regularly undertaken by owners or their servants, and recipes cover surface cleaning and varnish removal, as well as general guidance on care and protection of paintings (Rose 1998).

Protective coatings were frequently employed to prevent damage or deterioration. Paints and varnishes applied to wood and metals protected from the effects of water, and it is probable that, after polishing, Classical bronzes were coated with oil or a thin layer of bitumen (Born 1993, Craddock & Giumlia-Mair 1993). Evidence from excavated Anglo-Saxon swords indicate that scabbards were lined with sheepskin, complete with hair, and it is possible that this was used because inserting the sword into the fleece would have a polishing action and the natural grease (lanolin) would coat the sword-blade and protect it from corrosion. Coatings such as oils and lacquers may have been intended to have a dual role of enhancing the appearance as well as being protective, and they were probably renewed at regular intervals as part of a normal maintenance programme.

Insects have always caused serious damage to objects, and aromatic materials were used as repellants. Pliny comments on wood-boring insects and reports the treatment of papyrus with citrus oil to deter insects (Pliny the Elder XIII); strong-smelling woods such as cedar and herbs such as lavender have an apparently long tradition of being used for the storage of woollen textiles in order to repel clothes-moths. The damaging effects of light were also apparently well known. Campbell's study of the care of Netherlandish paintings in the fifteenth and sixteenth centuries indicates that large paintings were often protected by curtains and smaller ones with cupboards or boxes, and some with glass (Campbell 1998). Housekeeping in grand houses involved keeping furnishings clean and protecting them (particularly textiles) from the effects of light. Many of the techniques used in historic houses today (and in some museums) to limit exposure to light are similar to the traditional ones which are known to have been practised in great houses. Rooms or complete houses were closed for periods when the owners were not using them, and tapestries and other hangings, upholstered chairs, and carpets were often exposed only on special occasions and were safeguarded using covers, often of a similar colour but more robust and cheaper material, for everyday protection (Cornforth 1986).

Repair and restoration by craftsmen and women

Apart from domestic repair, damaged or deteriorated objects were repaired or restored by craftsmen and women whose main occupation was to make similar objects. These craftsmen were often highly skilled and had extensive knowledge of the working properties of their materials and the way they should be used for maximum effect and durability in a new object; they applied this knowledge when repairing old or damaged objects. Some accounts confirm that craftsmen not only created new objects but also repaired existing ones. Pliny discusses the restoration of ancient objects and monuments (Pliny the Elder XXXV), Theophilus (qv), writing in the eleventh century, mentions the cleaning of gilded surfaces and corroded metals, and Cennino Cennini, working in fifteenth-century Florence, included advice in his *Craftsman's Handbook* on glues for repairing specific materials including wood, paper and leather (Cennini V).

Paintings were restored by painters or sign-writers, furniture by joiners or cabinet-makers and buildings by masons, all of whom were working with the materials of their trade. This meant that repairs would be made with the same material, eg metal on metal, wood on wood, and the emphasis was on regaining or even improving the appearance and function of the object. Many of the painted objects subjected to restoration were religious images and here the emphasis was on maintaining the sacred image in an unblemished state (Daly Hartin 1990), to the extent of total repainting of panels and sculptures. In fact it is typical of much polychrome sculpture that it is covered in successive layers of repainting.

Exterior stonework is vulnerable to deterioration caused by dirt, pollution and salts. Protecting surfaces such as stone and plaster using a renewable coating of lime, sometimes with the addition of animal glue or tallow, is a traditional technique which was apparently used for aesthetic reasons and because it has slightly antifungal properties (Ashurst 1990). Some forms of limewashing continue to be used for the maintenance of ancient buildings and lime has been re-evaluated as a means of protecting and conserving decayed sculptural stone (Durnan 1997). There has recently been much discussion of the role of the calcium oxalate surface, known as *scialbatura*, found on some Roman and later marble and limestone structures. It is possible that this patina is, by chance, the end-product resulting from an original protective coating containing an animal protein as a binder which has subsequently been affected by the environment and by growth of microorganisms (Del Monte & Sabbioni 1987, Lazzarini & Salvadori 1989, Martin-Gil *et al* 1999).

Understanding of the working properties and deterioration behaviour of materials is shown in the care taken to use stable materials and methods of manufacture when making an object. This is seen both in the selection of pigments and construction of surfaces such as wood panels for paintings (Campbell 1998) and in the construction and maintenance of buildings. Traditional techniques show that there was a good understanding of how to design and maintain buildings, both for durability and to achieve an effective internal climate, a foreshadowing of preventive conservation. Ancient authors such as Vitruvius discussed selection of materials, positioning and orientation of buildings to gain maximum durability and to achieve the desired climate inside the structure (eg Vitruvius, Books 2 and 6). Later buildings also provide evidence of this sort of care over construction to achieve good ventilation and heating (Koller 1994). Similar approaches are now being emphasized in the construction of museums in order to establish an effective internal climate without resorting over-much to mechanical means such as air conditioning (Cassar 1995).

What clearly characterized both domestic and craft repair and maintenance was the requirement to keep the objects or structures in a condition suitable for use; again this is characteristic of modern conservation, though the use may be a museum use.

Emphasis on appearance: making things look good

Art-objects have a particular visual impact. Today their repair puts emphasis on maintaining or reconstructing the image, and, as far as it is understood, the appearance intended by the artist. Since the immediate visual effect has much to do with shape, colour and texture, the maintenance of the surfaces is a very important consideration. Because works of art have been valued and collected from very early times they have also been restored, and the precursors of modern conservators were the craftsmen and women who restored art objects for collectors and early museums. The aim was to ensure that the object still had visual impact so that it could be appreciated for its aesthetic qualities and remain in use as an art object, but it was not the kind of restoration practised today, since the original nature of the object and the artist's intentions were often disregarded.

The restoration of easel paintings is a good example. Just as other types of object were repaired by craftsmen accustomed to create the same types of object, so paintings were often restored by painters using the materials of their trade. Throughout the sixteenth to the nineteenth centuries travelling painters cleaned, retouched (repainted damaged, or sometimes undamaged, areas) and revarnished pictures, and some made a living by specializing in restoration. In the eighteenth century the Spanish Royal collection was the subject of a programme of restoration, after a disastrous fire in 1734, and the records reveal that of the ten people employed only three were considered to be restorers while the remaining seven were painters (Veliz 1998).

The processes used concentrated on maintaining or improving the image, and some restorers might choose a treatment because it demonstrated their skills and expertise rather than because the condition of the painting required it (Daly Hartin 1990). Cleaning involved removal of obscuring 'dirt' from the surface or, more drastically, removal and replacement of a dirty or discoloured varnish. Manuals and other records of practice, particularly from the eighteenth and nineteenth centuries, show that there was much discussion of cleaning materials, and of the quality and colour of varnishes, though some practitioners preferred to keep their recipes secret. Removal of varnish would, almost inevitably, lead to removal of old repairs and the need to repaint the damaged areas to restore the unity of the picture (Brommelle 1956, Sitwell 1998, Talley 1998). 'Before the twentieth century retouching had a single simple objective: to make a painting look as good as possible in accordance with the wishes of the owner and restorer' (Bomford 1994b, 34).

Extensive repainting could be undertaken and the original image was not always adhered to. In fact paintings might be considerably altered, as was revealed during the restoration of Peake the Elder's portrait of Henry, Prince of Wales (probably painted about 1610), in which the figure of Time with his scythe had been painted over, possibly thirty or more years later, thus altering both the balance and meaning of the picture (McClure 1992). Major artists such as Van Dyck and Reynolds are known to have undertaken restoration. Reynolds was apparently very proud of his abilities as a restorer but he was not above making considerable alterations to suit his views, such as substituting a light sky for the original dark background in a Velasquez painting (Talley 1998). Although, by the late eighteenth century, the problems of retouching were being widely discussed (such as the durability of materials, the choice of appropriate textures and the desirability of not covering original paint), restoration continued during the nineteenth century in much the same way as before, frequently paying little respect to the original (Bomford 1994b).

Works of art on paper fared in much the same way as paintings. Publications provided advice on methods and materials for cleaning and repair. Drawings were often heavily restored, some by artists such as Rubens. The manufacture of prints provided the possibilities of making further impressions at a later date. During the nineteenth century, original copper plates by earlier artists were sometimes used to produce new impressions, not always to deceive, but some were apparently passed off as originals (Donnithorne 1988, Kosek 1994).

Restoration of buildings

The restoration of buildings was approached during the nineteenth century in much the same way as the restoration of works of art. Many major monuments throughout Europe were in a state of disrepair at the beginning of the century (indeed ruins had been much admired due to the influence of the Romantic movement of the early nineteenth century), so there was scope for widespread reconstruction and restoration. Buildings were repaired to maintain their function but often considerably altered or rebuilt to suit the restoring architect's concept of the style of the original, rather than by following the design and detail of the remaining structure (Bristow 1990, Lasko 1992). This type of restoration is epitomized by the treatment given in the 1840s to the well-known round Norman church of the Holy Sepulchre in Cambridge, UK, which involved removing the fifteenth-century additions to the tower, replacement of perpendicular windows with windows in the Norman style, rebuilding part of the wall and vault, re-roofing, and removal of the later furnishings. The result was a marked change in appearance and an obvious loss of original features (Bristow 1990).

One of the foremost exponents of this form of restoration was the French architect Eugene-Emmanuel Viollet-le-Duc who exercised considerable freedom in deciding on the appropriate finished state while restoring monuments such as the cathedral of Notre Dame in Paris or the mediaeval town of Carcassonne (Lasko 1992). Since this restoration sanctioned the removal of later additions and their replacement with features in 'the original style', there was widespread loss of historical evidence and of character. Nevertheless many buildings which had been in very poor condition had been brought back into use.

Development of museum restoration

As collections grew and developed from personal possessions to public museums, so did the role of museum restoration become more clearly identified. At first, of course, owners of collections cared for them themselves. The seventeenth-century collector Sir Hans Sloane (whose collection eventually formed the basis of the British Museum) is recorded as remarking on the poor condition of a collection of natural and ethnographic objects he had just acquired and commenting, 'I found myself obliged to take immediate Care of all of them, and in the first place the Animal substances , which are most subject to Destruction' (quoted by Alexander 1983, 29). A portrait of Charles Wilson Peale, the Director of the eighteenth-century Philadelphia Museum, shows him not only with the contents of his museum in the background but with the tools he used to restore natural history specimens and paintings laid out in the foreground (Alexander 1983).

During the seventeenth, eighteenth and much of the nineteenth centuries objects in collections were restored by independent craftsmen working in the traditional way. The emphasis was on maintaining the appearance of objects and could involve techniques such as reshaping crushed objects, removing surface discoloration or corrosion, replacing crumbling material, and repainting to conceal damaged areas. No records were kept of materials or techniques. However, with the creation, in the eighteenth and nineteenth centuries, of the great public museums such as the Louvre and the British Museum, and national galleries of art, came the development of public interest in the welfare of national collections. Members of the public began to criticize activities in museums: for example, the Marin affair in 1797 during which the restoration of paintings in the Louvre was questioned (Massing 1998), and the controversy surrounding methods of cleaning paintings belonging to the National Gallery in London which led to a Committee of Enquiry of 1853 (Brommelle 1956). Such public discussion alerted the authorities and the restorers to the need to re-evaluate restoration practices, and to give a proper account of methods and materials.

Gradually major institutions began to employ restorers to work on their collections. For example, it is known that John Doubleday was employed at the British Museum to work on the ivories from Nineveh (Seeley 1987) as well as restoring the famous Roman cameo glass Portland Vase in 1845 (Fig 6) (Williams 1989), and that Robert Ready was employed in the 1850s to restore pottery and to make replicas of coins and medals, and was later joined by his sons Talbot and Augustus (Oddy 1992b, Shorer 1997). These restorers would have been responsible for training their successors and passing on knowledge not only of techniques and materials but of the objects themselves.

Figure 6 John Doubleday after restoring the Portland Vase at the British Museum in 1845. Beside him is the vase itself and a picture of the fragments before restoration (© Copyright The British Museum).

The concept of preventive conservation began to emerge in the second half of the nineteenth century when mechanisms of deterioration began to interest scientists. Before this period objects were protected, but not always with the needs of conservation in mind. Many early collections had elaborate cases and drawers built for them but these were intended to promote the organization and display of the collection rather than to protect the objects from the environment (well illustrated in Hooper Greenhill 1992). Small objects were often protected in boxes; for example Sir Hans Sloane used specially-made boxes for insect specimens designed to allow the specimens to be examined but also to protect them from damage by other insects (Alexander 1983). It was also customary to use glazed wooden display cases during the nineteenth century, which would certainly have been protective against extremes of environment, even though they were used more for reasons of security than conservation.

The nineteenth century was a period of rapid industrial and urban growth and of consequent air pollution caused by coal-burning and gas-lighting. It was investigation of the effects of air pollution which prompted the development of ideas on preventive conservation. The problem is clearly illustrated in the vivid description of the conditions at the National Gallery in London in the 1850s. The Gallery was regularly very crowded and 'when the atmosphere grew heavy the windows were opened and the smoke from the nearby municipal wash-house, which was described as "very opaque" came in through the windows which were not protected with gauze' (Brommelle 1956, 177). It is not surprising, then, that there was a thick 'greasy' film on the paintings and the rate of deterioration was marked. The cleaning of the affected paintings was controversial, and the subsequent public enquiry prompted the examination of methods for preventing the damage caused by pollution, as well as evaluation of different approaches to cleaning.

The nineteenth century was also a period of energetic scientific enquiry. However, despite a growing interest in the deterioration of museum objects and works of art, traditional approaches to restoration continued throughout the nineteenth and well into the twentieth century. During the 1853 enquiry into the cleaning of paintings at the National Gallery, London, the question of employing a chemist at the Gallery was raised, and was supported by the distinguished scientist Michael Faraday, but Sir Charles Eastlake, soon to become the first Director of the Gallery, was against a permanent appointment and a chemist did not join the Gallery staff until much later (Brommelle 1956). However, in 1888 concern for the condition of the collections in the Königlichen Museen, Berlin, led not just to an enquiry but to the first permanent appointment of a scientist to a museum position, focusing on the investigation and restoration of objects. Friedrich Rathgen became the Director of the specially established Chemical Laboratory in Berlin and remained in this post until 1927. He was certainly the first to adopt a systematically scientific approach to devising and perfecting conservation treatments. His pioneering work, and his publication of his methods (Rathgen 1905) had a profound effect on the future development of archaeological and museum conservation (Gilberg 1987, Plenderleith 1998).

In the UK a similar laboratory was established at the British Museum immediately following the First World War. Many objects had been stored in the London underground railway system to protect them from possible war damage, but the damp conditions had caused extensive deterioration. It was clear to the Trustees of the Museum that the problem was beyond the abilities and techniques of the Museum's existing restorers, and that emergency measures were needed. In 1920 a temporary laboratory was set up under a chemist, Alexander Scott, in the hope that the problem could be dealt with in three years. The need for a permanent conservation laboratory eventually became apparent to the authorities, and in 1931 it was formally established as the British Museum Research Laboratory. Harold Plenderleith had joined the laboratory in 1924 and was instrumental in ensuring that the collections were housed in much better conditions during the Second World War. In fact the effects of the first war-time storage, and planning for the second period of storage played an important role in establishing optimum conditions for preventive conservation, particularly for avoidance of mould growth (Oddy 1997, Plenderleith 1998). The work of Scott and Plenderleith was, like Rathgen's, characterized by the systematic application of scientific method to developing and refining conservation procedures.

Despite the involvement of chemists in museum conservation in the early years of the twentieth century, most objects conservators continued to be drawn from a craft background, and were trained on the job. Taking the British Museum as an example, during the 1920s and 1930s several 'boy-learners' were trained both at the Central School of Arts and Crafts and in the British Museum Research Laboratory, and worked as members of the conservation staff for forty to fifty years (Shorer 1997, Plenderleith 1998). This situation changed only gradually until the last quarter of the century when the number of science-based university conservation training courses and thus the number of specialist conservators increased. Manual skills and a knowledge of the properties and behaviour of materials gained from learning a craft such as pottery or metal-working are still essential for anyone training to be a conservator, but are now normally combined with a science training. The craft tradition continues to be strong in some areas of conservation, particularly in the conservation of buildings where skilled craftsmen or women carry out the work on the structure, normally under the direction of a conservation architect. In this case many of the traditional skills, almost lost during the second half of the twentieth century, have been deliberately reinvigorated.

The restorers working under the guidance of scientists such as Plenderleith were still involved in rectifying damage and maintaining the appearance of objects, but they were now also involved in attempting to halt or slow deterioration, as well as to prevent it in the future. They seldom had a scientific education but they were highly skilled and had an extensive knowledge of raw materials and traditional technologies, as well as an impressive familiarity with the collections they were working on, often built up over many years of service within the museum.

Care of archaeological objects

The science developing during the nineteenth century involved identification and classification of natural phenomena. During the same period similar processes of collection and classification were applied to the material remains of the human past in an attempt to move beyond the sometimes romantic notions of antiquarianism to a clearer understanding of antiquity: this was the beginning of the science of archaeology. In 1807 the Danish government created a Royal Committee for the Preservation and Collection of National Antiquities which was to be responsible for creating a museum, and protecting field monuments. In 1816 Christian Jurgensen Thomsen became the secretary to the committee and at the same time the first curator of the National Museum. Thomsen was faced by a large and growing collection of antiquities which had accumulated without any organization, and he is well known for developing the concept of the three-age system (Stone Age, Bronze Age, Iron Age) which was first suggested by Vedel-Simonsen, and for using this as a basis for classifying and displaying the material in the museum (Daniel 1975, Schnapp 1996). What is not so widely known is that Thomsen also promoted the care and treatment of antiquities. When he took responsibility for the collections they were both disorganized and very dirty and he set out to repair them as well as to organize them. He was also concerned with the problem of freshly excavated objects, particularly waterlogged organics, fragile pottery and corroding iron, and well aware that excavation might induce very rapid deterioration. His letters and other writings contain advice on effective first-aid treatments to be applied to objects immediately after excavation (Brinch Madsen 1987, Jakobsen 1988). What is notable about Thomsen's approach to conservation is that he was repairing objects not just to form an archive, but so that he could use them in the development of the discipline of archaeology.

Thomsen is often considered to be the originator of modern archaeology (for example Daniel's *A Hundred and Fifty Years of Archaeology* (1975) was measured from the date of the opening of the Danish National Museum displaying the three-age classification). Although working largely empirically he can also be seen as one of the key figures in the development of modern archaeological conservation. Through his assistant C F Herbst and subsequently through Georg Rosenberg, who joined the Museum in the 1890s and continued to work there until his death in 1941, the practice of what certainly would have been recognizable as archaeological conservation was firmly established at the National Museum (Brorson Christensen 1970).

Herbst developed the alum treatment for waterlogged wood in 1859, and this was used successfully to preserve many objects excavated during the nineteenth century which would otherwise have collapsed and disintegrated beyond recognition. Very little is known about how he developed this treatment, or what else he tried before arriving at his particular method. He obviously understood the porosity and instability of waterlogged wood from first-hand experience, and must have sought a material which would enter and harden in the pores. Alum was traditionally and widely used as a mordant in dyeing textile fibres so its general properties were well-established, but he may well have taken scientific advice before deciding to test it (Brorson Christensen 1970, Brinch Madsen 1987). That ideas on the care and conservation of antiquities were well-advanced in Denmark is also reflected in the fact that the first written guidance on the handling of archaeological objects was produced there by M R Schmidt in 1858 (Sease 1996). The development of archaeological conservation represented a shift from restoration (focused on maintaining the visible character of an object) towards an acknowledgement of the importance of conserving the evidence of past culture, not all of which is immediately apparent when viewing an object.

Interest in antiquities was also well advanced in Germany, as seen in the large number of investigations of metal corrosion and patina cited by Rathgen in his own book (Rathgen 1905), and in the publication by Albert Voss of the *Merkbuch Altertümer Aufzugraben und Aufzubewahren* in Berlin in 1888. This was a small handbook on the excavation and care of objects which was popular enough to run to two later editions in 1894 and 1914 (Gilberg 1987). Without this interest in archaeological material in Germany it is unlikely that there would have been an enquiry into the state of the collections or the foundation of a laboratory in Berlin.

Some archaeologists, such as Schliemann and Petrie, undertook the treatment of the objects they excavated, and Petrie, in particular, was clearly committed to the care of the material he brought back to London from his excavations. In 1888 he wrote a short paper entitled 'The treatment of small antiquities' for the *Archaeological Journal* which was largely about the cleaning of corroded metal objects (Petrie 1888). Lucas, a chemist, who worked in Cairo during the first half of the twentieth century, undertook a number of technological investigations of ancient Egyptian objects, and, while working for Howard Carter, was responsible for the conservation of much of the material from Tutankhamen's tomb (Gilberg 1997). Archaeological objects presented new problems for which an effective approach did not exist. Classical statuary and well-fired ceramics had survived reasonably well but excavated metals and organics might be unrecognizable or were likely to deteriorate rapidly and devastatingly. They could not be saved by restorers using traditional repair methods, so it was natural to look toward scientists for advice and suggestions, and to try out chemical treatments.

Development of materials science

With the definition and growth of science in the nineteenth century scientists became involved in the study of deterioration of works of art. Thus Sir Humphry Davy, the distinguished English chemist, having analysed some of the pigments from the wall paintings at Pompeii (Rees-Jones 1990, Roy 1998), was asked in 1818 to advise on the papyri from Herculaneum and made a study of their condition before suggesting ways (unfortunately unsuccessful) of improving on the current unrolling technique (Gilberg 1988a). There were numerous studies of corrosion on metals including an investigation of bronze corrosion by Humphry Davy's brother John in 1826 (Seeley 1987). Rathgen (1905) cited 23 papers, mostly in German, published between 1879 and 1904 on the corrosion of iron, four of them about iron antiquities, and another 28 papers written between 1853 and 1903 on the investigation of patina on bronze antiquities. Surprisingly, he did not apparently know of the 1882 paper on the corrosion of archaeological iron by Edward Krause, a chemist working in Berlin, in which Krause identified chlorides and sulphates in the corrosion layers (Jakobsen 1988) but he did cite the 1894 paper by Marcellin Berthelot who investigated the deterioration of Egyptian bronzes in the Louvre and identified the electrochemical nature of the corrosion known as bronze-disease (Gilberg 1988b).

The deterioration of paintings also attracted attention. In 1850 Michael Faraday, one of the greatest British scientists of the nineteenth century, was asked to join a three-man committee charged with investigating the state of the paintings in the National Gallery in London (Brommelle 1956), and subsequently undertook some experiments on the effect of varnishes as protective barriers on paintings. In 1888 W J Russell and W de W Abney (both Fellows of the Royal Society) were invited to undertake a study of the effect of light on watercolour paintings after a public controversy over the proposal that the Gallery should be open in the evenings, which would thus expose the paintings to the unknown hazards of artificial (gas) light. Their report emphasized the role of the blue and violet end of the light spectrum in the fading of pigments (Brommelle 1964). Scientific investigation of pigments and of the structure of paintings, including the use of cross-sections, developed in the early years of the twentieth century (Roy 1998).

Other ancient materials, such as glass, also attracted interest. In 1861 Sir David Brewster, a physicist noted for his work on light and the refractive index, published the results of his study of the surface effects of iridescence on ancient glass, and in 1880 James Fowler published a scientific study of the different forms of decay found on glass antiquities (Seeley 1987, Sease 1996).

Although these pieces of research were individual events, together they provided the basis for the development of a wholly new scientific approach to the study of objects. The products of deterioration were investigated, and some researchers linked the type of deterioration to the burial environment from which the objects came (eg the work of von Fellenberg on bronzes, published in 1865 and cited by Rathgen). This was the birth of materials science, which altered the focus of the study of artefacts from surfaces and appearances to structures and mechanisms of change, and thus encouraged the application of new examination techniques such as X-radiography, thin-sectioning of paint layers, and analysis of pigments and corrosion products.

An understanding of raw materials and deterioration was an essential prelude to deciding if changes could be halted or minimized (rather than repairing or improving the appearance, as most restorers had until then). However, this interest in investigation did not bring about a major change in treatment methods, rather it may have contributed to a change in emphasis from keeping things in use to treating them as the passive subjects of research.

Emergence of conservation

Although there is evidence for the general maintenance of ancient objects and buildings, and the deliberate care of some with the intention of preserving them for the future 'there are good reasons for believing that a sense of responsibility toward artistic heritage emerged only slowly' (Marijnissen 1996, 279). Conservation as it is known today did not develop until the 1920s to 1930s, and the twentieth century was characterized by the definition and establishment of conservation, by the founding of conservation organizations and the codification of ethics and standards of practice. Although the term conservation was not new, for example the Royal Institute of British Architects had produced a booklet in 1865 entitled *Conservation of Ancient Monuments and Remains* (RIBA 1865, Jokilehto 1999), it was deliberately chosen, in preference to preservation or restoration, to describe the activities of the newly emerging discipline. Conservation was preferred because it embodied the need to protect the integrity of objects including the evidence of their history (this is clearly set out in the objectives of the International Institute for Conservation drawn up in 1950: IIC 1999).

A major event in this development was the International Conference held in Rome in 1930 under the auspices of the League of Nations. The conference had been called to discuss the standards of restoration of paintings and the results of the deliberations were published in 1939 as the *Manual of the Conservation and Restoration of Paintings* edited by Harold Plenderleith, George Stout and Helmut Ruhemann. Not only was this one of the first assemblies of specialists involved in conservation, but the resulting publication indicated by its title that conservation was an activity distinct from, though closely related to, restoration.

In 1931 there was a similar International Congress in Athens focusing on the conservation of monuments. Its conclusions were set out in the *Athens Charter* which embodied concepts such as the need to respect the evidence of the history and development of the monument (Jokilehto 1999). A further landmark in the evolution of conservation came immediately after the Second World War with the founding of the International Institute for Conservation of Historic and Artistic Works (IIC). Discussion had started in the USA in 1946, and the Institute was formally founded, in London, in 1950 (IIC 1997, Plenderleith 1998). Over the next thirty to forty years local groups of IIC were founded and then themselves developed into separate national conservation organizations, such as the American Institute for Conservation and the UK Institute for Conservation.

Whereas restorers were often secretive about their methods, the involvement of scientists in conservation was probably responsible for introducing the concept of producing written reports on the work undertaken, now, of course, considered an essential aspect of professional practice. *The Museums Journal* (the journal of the UK Museums Association) was founded in 1901 and before the advent of specialist conservation journals articles on the care and conservation of objects were published in it; several came from authors outside the UK since there was, as yet, no international journal. Both Rathgen's book *The Preservation of Antiquities* (1905) and Scott's third report *The Cleaning and Restoration of Museum Exhibits* (1926) provided detailed accounts of the methods and materials they used and Plenderleith produced a number of reports and short books including *The Preservation of Antiquities* in 1934. But it was his comprehensive book, *The Conservation of Antiquities and Works of Art*, published in 1956 and translated into five other languages, which had a profound effect on the development of objects conservation, particularly conservation of archaeological objects.

Plenderleith's choice of title again confirmed the existence of the new discipline of conservation. Equally influential was (and is) the journal *Studies in Conservation*, introduced by IIC in 1952, though its early editions clearly show that the investigation and conservation of paintings were the major preoccupations in the discipline, with relatively few papers on other types of object. This, perhaps, reflects the fact that *Studies in Conservation* was initially introduced to take the place of *Technical Studies in the Field of the Fine Arts*, a short-lived but pioneering publication on the investigation and conservation of paintings produced by the Fogg Art Museum, USA, between 1932 and 1942 (IIC 1997, Plenderleith 1998).

Until the 1920s the only laboratories dedicated to the conservation of antiquities were the laboratory in Copenhagen, Rathgen's laboratory in Berlin and the Textile Laboratory which was set up in Stockholm in 1908 under Agnes Branting and which pioneered the conservation of ancient textiles (Geijer & Franzen 1975). The 1920s and particularly the 1930s were a time of enormous growth and activity, with an increase in national awareness of conservation. Laboratories and research departments were created in major museums and galleries across the world. Amongst the earliest was the research department at the Fogg Art Museum, Harvard University, set up in 1925 by Edward Forbes, for the study of paintings and their conservation. At about the same time a laboratory was set up in Cairo under Alfred Lucas (Gilberg 1997). The laboratory at the Louvre was set up in 1931, and no less than three laboratories, at the Courtauld Institute, University of London, at the National Gallery in London and at the Musées de Belgique in Brussels, were set up in 1934. The Doerner Institute was founded in Munich, and a laboratory was established in the new Institute of Archaeology, University of London, both in 1937, laboratories were set up in New Delhi and in Tokyo in 1938, and the Istituto Centrale del Restauro was founded in Rome in 1939 (Plenderleith 1998, Roy 1998, Jokilehto 1999).

The damage to cultural material during the Second World War provided impetus for the foundation of a number of important international organizations concerned with the cultural heritage, all of which have had an effect on the development of conservation at an international level. The International Council for Museums (ICOM) was founded in 1946 and the United Nations Educational, Scientific and Cultural Organization (UNESCO) was initiated in the same year. Under the auspices of UNESCO the International Centre for the Study of the Preservation and Restoration of Cultural Property, soon known as 'the Rome Centre', was founded in 1959, with Plenderleith as its first Director (the Centre is now known as ICCROM, the International Centre for Conservation in Rome). This Centre has played an important role in promoting conservation through providing advice internationally, offering training courses, and maintaining a specialist library. The International Council for Monuments and Sites (ICOMOS) was founded in 1965, and two existing committees were amalgamated in 1966 to form the ICOM Committee for Conservation (ICOM-CC), which again has had a profound effect on the development of conservation through its international conferences and meetings.

The first half of the twentieth century was, thus, characterized by the definition of conservation and its inclusion in the activities of both national and international organizations. The second half of the century was characterized by growth and development of formal training courses, formulation of codes of ethics and standards for the practice of conservation, and by the tackling of issues such as professional accreditation. It also saw, certainly in the UK, the establishment of conservation posts in smaller, non-national museums in the 1970s. This was followed by a contraction of activities in the late 1980s and 1990s, resulting in many conservators leaving museums to work as independent specialists, now employed on short contracts to deal with individual museum projects. Much of the growth in the 1970s was linked, not just to an increasing awareness of conservation, but to the growth of rescue or salvage archaeology. The considerable increase in excavations, and in finds needing conservation (Swain 1998) encouraged the creation both of conservation posts, and of training courses intended to produce conservators to fill them.

In terms of conservation practice, the twentieth century saw an increasingly scientific approach, and the harnessing of scientific developments from other fields. The emerging discipline of conservation was characterized by the acceptance that a knowledge of the materials from which objects were made, and of the mechanisms of their deterioration, as well as an understanding of the scientific basis for the design and use of treatments, were essential. X-radiography was adopted to examine the internal structure of objects, and the invention of methods of examination such as scanning electron microscopy and of instrumental analysis such as X-ray diffraction enhanced greatly the possibilities of understanding materials and their deterioration. By the end of the century scientists working on ancient materials had adopted a wide range of analytical techniques (Tite 1972, Bowman 1991, Lambert 1997).

The industrial development of synthetic polymers (plastics) also had a marked effect on conservation practice. Until the middle of the century restorers and conservators had been largely dependent on natural adhesives such as animal glues, starch pastes and beeswax, all of which form coloured or opaque films, whereas many synthetic polymers form more-or-less transparent films. The first semi-synthetic material, cellulose nitrate (then known as gun-cotton), had been used by Rathgen and by archive restorers in the final years of the nineteenth century (Rathgen 1905), but it was not until the introduction of fully synthetic polymers such as polyvinyl acetate that the possibilities of these materials for conservation were realized. During the second half of the twentieth century these materials were adopted throughout the field of conservation. For example the discovery and raising of the waterlogged warship, the *Wasa*, in 1970 from Stockholm Harbour provided the impetus to research and test the use of a synthetic polymer, polyethylene glycol, which eventually revolutionized the treatment of waterlogged wood (Pearson 1987) and made it possible to treat whole ships and boats successfully. However, synthetic polymers have their disadvantages and the enthusiastic adoption of these new materials was followed by a realization of their limitations and the evolution of a more cautious approach to their use. The final years of the century saw a deliberate re-evaluation of traditional natural materials, particularly in the field of conservation of organic materials, and this included the testing of sucrose (sugar) as an alternative to polyethylene glycol in the treatment of waterlogged wood, and starches as adhesives for textiles and ethnographic basketry (Brommelle & Pye 1984, Horie 1987, Cruickshank & Tinker 1995).

Conservation principles

The principles of conservation established during the twentieth century began to emerge during the nineteenth century. A major influence was the reaction against what were seen as the excesses of restoration of paintings and of buildings, which encouraged the formulation of ideas about more appropriate approaches to understanding and preserving works of art and monuments.

The seeds of these ideas can be seen in publications such as Bonnardot's paper on the art of restoring prints, published in 1846, which pointed out that stains might be historical evidence which should not be removed, and in Secco Suardo's book on the restoration of paintings, published in 1866, which presented a cautious view of methods and materials of retouching (Bomford 1994b, Kosek 1994). In 1869 a German restorer, Pecht, was beginning to advocate what would now be called minimum intervention by suggesting that 'it would be better to accept the damage on a painting rather than to disguise it' (Daly Hartin 1990, 33).

Perhaps even more influential in the development of modern thinking about conservation was the outcry against the nineteenth-century destruction of buildings, led by John Ruskin in the UK. In 1849 Ruskin published *The Seven Lamps of Architecture* and the chapter entitled 'The Lamp of Memory' embodies much of his passionate feeling for old buildings and his equally passionate repudiation of the then current form of restoration. Of an old building he said: 'Its glory is in its Age, and in that deep sense of voicefulness … which we feel in walls that have been washed by the passing waves of humanity'. His view was that this history embodied in a building 'connects forgotten and following ages' and gives the building the gifts 'of language and of life' (Ruskin 1883, 186–7).

Of restoration he said:

> *Neither by the public, nor by those who have the care of public monuments, is the true meaning of the word* restoration *understood. It means the most total destruction that a building can suffer. (Ruskin 1883, 194)*

> *Do not let us talk then of restoration. The thing is a Lie from beginning to end. You may make a model of a building as you may of a corpse, and your model may have the shell of the old walls within it as your cast might have the skeleton ... but the old building is destroyed, and that more totally and mercilessly than if it had sunk into a heap of dust, or melted into a mass of clay: more has been gleaned out of desolated Nineveh than ever will be out of re-built Milan. (Ruskin 1883, 196)*

He went on to say:

> *it is again no question of expediency or feeling whether we shall preserve the buildings of past times or not.* We have no right whatever to touch them. *They are not ours. They belong partly to those who built them, and partly to all the generations of mankind who are to follow us. (Ruskin 1883, 197) (Note that the emphasis is Ruskin's.)*

Ruskin's views were accepted only slowly and restoration continued throughout the century in the UK. For example, towards the end of the century the West Front of St Alban's Cathedral, including a fifteenth-century window, was demolished and rebuilt according to the designs of Lord Grimthorpe, who had also designed the Westminster clock now know as Big Ben (*Encyclopaedia Britannica* 1911, Lasko 1992). However, in 1877, when Sir George Gilbert Scott, one of the foremost architects of the day in the UK (and designer of the Albert Memorial), threatened to remove the seventeenth-century furnishings from Tewkesbury Abbey, the designer and craftsman, William Morris, decided to found the Society for the Protection of Ancient Buildings (SPAB) (Bristow 1990).

The principles set down in the *Manifesto* which was drawn up when SPAB was founded provide a further milestone in the thinking which has shaped conservation. They argue for daily care, visible repairs and minimal interference, rather than what was seen as the destruction wrought by nineteenth-century restoration. The final paragraphs read:

> *It is for all these buildings, therefore, of all times and styles, that we plead, and call upon those who have to deal with them to put protection in the place of restoration, to stave off decay by daily care, to prop a perilous wall or mend a leaky roof by such means as are obviously meant for support or covering, and show no pretence of other art, and otherwise to resist all tampering with either the fabric or ornament of the building as it stands; if it has become inconvenient for its present use, to raise another building rather than alter or enlarge the old one; in fine to treat our ancient buildings as monuments of a bygone art, created by bygone manners, that modern art cannot meddle with without destroying ...*
>
> *thus, and thus only, can we protect our ancient buildings, and hand them down instructive and venerable to those that come after us. (SPAB 1903, 74–5)*

At about the time that SPAB was founded, J J Stevenson, an architect who had trained under Sir George Gilbert Scott, gave a lecture to the Royal Institute of British Architects (RIBA) entitled 'Architectural Restoration, its Principles and Practice', in which he set out two important principles: firstly that buildings were historical records, and secondly that even relatively recent additions should be respected (Lasko 1992).

Both Ruskin's and Morris's ideas had a considerable effect on restoration and on the protection of buildings in mainland Europe and beyond. In 1879 a proposed restoration of the west front of St Marks in Venice was halted. In France an association was founded with similar aims to SPAB, known as *L'Ami des Monuments*, and in 1887 a law was passed for the protection of ancient buildings, which in turn influenced countries such as Algeria and Tunisia. In Germany a conservation journal, *Die Denkmalpflege*, was founded in 1899, and by the beginning of the twentieth century annual conservation meetings were being held (Jokilehto 1999).

It is clear that ideas were forming on the concept of a common heritage and the requirement to safeguard it, on the nature of the object and the information it may hold, including the visible signs of age (the object as a document), on using preventive care in preference to interventive restoration, on limiting the restorer's intervention only to what was really necessary, and on making such interventions visible. All of these have become important features of modern conservation philosophy.

What had also developed in reaction to what was seen as the damage caused by restoration was a warning against anything that could be construed as interference, in other words the injunction was 'hands-off!' This is epitomized by Ruskin's saying of ancient buildings that 'we have no right to touch them' and SPAB's recommendation that if a building becomes inconvenient for its use a new building should be erected rather than altering the original in any way. These views were understandable given the time and context, but they became untenable. In fact SPAB admitted as much in 1924 and added a note to its *Manifesto* which indicated that 'a modest addition' to an ancient building would be allowable provided it was built in modern style, was necessary and permanent and in no way attempted restoration. Today, concepts such as 'adaptive re-use' and urban renewal are essential tools for keeping old buildings alive. On the other hand, where objects are concerned there is sometimes a reluctance among archaeological and museum conservators to allow their use. This may partly stem from the caution engendered by the warning not to interfere, and perhaps from the tendency to treat objects as repositories of evidence to be protected in an archive, rather than as tools to be used actively in research and interpretation.

Study of early conservation materials

A wide range of different 'early' adhesives and gap-filling materials can be found on ceramics and other types of artefact in museum and other collections. Some objects were repaired domestically or by itinerant handymen, some by highly skilled museum restorers, but before the advent of synthetic polymers in the 1930s different combinations of a basic range of organic and inorganic materials were relied upon. Routine documentation of restoration and conservation methods and materials was not reliably established until, perhaps, the 1970s, so conservators working today must have a reasonable knowledge of what materials might be present in an object as a result of earlier treatments. The study and recognition of these materials is important for several reasons: some of them are probably worth reassessing since modern synthetic polymers have disadvantages; some old repairs should be preserved as part of the history of the object itself, as well as a contribution to the history of conservation; some may be disfiguring or causing damage; a broad idea of what materials or techniques may have been used can be helpful in deciding on approaches to testing and removal; some materials may have 'unexpected' ingredients which may hinder the interpretation of an analysis; and, finally, some may be toxic.

Early conservation materials include the obvious such as animal glues, waxes and plasters, as well as the less known such as sulphur, blood protein, zinc powder or gutta percha. They have been very effectively catalogued in papers such as those by Hanna and Lee (1988), Koob (1998), Thornton (1998) and Williams (1988). The authors have drawn on the evidence in the objects themselves, as well as information given in early handbooks such as *Spon's Workshop Receipts*: the 1909 edition of Spon has 50 double-columned pages devoted to 'cements and lutes' with extensive recipes (Spon 1909).

The introduction of synthetic polymers had a profound effect on the development of conservation and chronicling the early use of these materials is equally important. Howie's paper on the materials used on fossil specimens since 1930 provides a picture of the impact of synthetic polymers on the conservation of weakened and fragmentary materials such as fossils and archaeological objects (Howie 1984). The first synthetic to be used on a fossil specimen was Bakelite in 1928 and vinyl polymers were introduced in the 1930s and 1940s but (not surprisingly) many of the traditional materials continued in use. Testing and adoption of synthetic polymers was, in fact, slow until the 1960s; nevertheless, curators and conservators need to be aware that they may find a synthetic polymer on an object treated well over half a century ago. Accounts of materials in use since that time make an important contribution to the history of conservation as well as providing information to the conservator dealing with previously treated objects. Examples include the survey of adhesives and consolidants used on glass (Davison 1984) and other papers given at the IIC Paris Congress on Adhesives and Consolidants (Brommelle & Pye 1984), Horie's book on adhesives used in conservation (Horie 1987) and a recent edition of the *Journal of the American Institute for Conservation* which contains nine papers on filling materials, including those by Koob and Thornton, already cited (JAIC **37**(1) 1998).

Evolution of preventive conservation

An important development during the second half of the twentieth century was the approach to conservation, initially called Museum Climatology or 'control of the museum environment', now called 'preventive conservation'. The realization that the environment in which objects were kept (light, temperature, relative humidity, pollution etc) contributed to deterioration led to the adoption of techniques for monitoring and control. The growing interest, during the nineteenth century, in the condition of collections (eg of the paintings in the National Gallery, London) and the involvement of scientists in the study of materials and their deterioration, had begun the process of looking for *causes* of deterioration from which developed the concept of *preventing* deterioration. Rathgen had commented on the need to protect objects from the effects of light (Rathgen 1905), and the damage caused to the British Museum Collections stored in damp conditions during the First World War led Plenderleith to evaluate the effects of humidity on objects, which led directly to much better storage conditions for the collections during the Second World War (Caygill 1993, Plenderleith 1998).

This branch of conservation was established rapidly in the 1950s and 1960s. The first chapter in Plenderleith's book *The Conservation of Antiquities and Works of Art*, published in 1956, was entitled 'The Influence of the Environment', clearly showing his interest in environmental causes of deterioration and in methods of controlling them. In 1961 the International Institute for Conservation (IIC) held a conference in Rome, entitled *Recent Advances in Conservation*, during which twelve of the forty-seven papers were on the subject of the museum climate and on the effects and control of insect and fungal damage (Thomson 1963). In 1964 the Murray Pease report, which was one of the first attempts to set out standards for practice, included some broad guidelines for safe handling and good storage conditions as means of protecting objects from accidental damage. By the late 1960s the concept that the 'museum climate' could be regulated in order to prevent some aspects of deterioration was fully accepted and IIC dedicated a one-week conference in London to the subject of *Museum Climatology* (Thomson 1968).

From the 1960s onwards several factors may have prompted the increasing emphasis on preventive measures based on control of climate and activities such as handling. One was the growing understanding of the damage caused by pollution to human health, to wildlife and to stone in buildings and monuments. A second was the realization that modern synthetic polymers which had seemed to present major advantages for the development of conservation treatments could not only deteriorate themselves but cause major damage to objects, and a third may have been the continuing controversies over restoration of paintings. The guiding principle of minimum intervention, which had arisen out of the requirement to protect the integrity of objects, was probably also responsible for shifting the emphasis of conservation towards developing preventive measures which did not involve direct interaction with objects.

By the late twentieth century it became apparent that many collections were in poor condition and that some form of integrated management was needed to establish more effective preventive care of collections. From this developed the concept of managing the use and welfare of collections in museums, now usually known as collections management.

Conservation at the beginning of the twenty-first century

The last quarter of the twentieth century was a time of considerable change in conservation, characterized by the development of collections management and environmental management, and the increasing emphasis on preventing deterioration. The adoption of a wide-ranging, holistic view of what constitutes conservation and collections care provided management frameworks within which a range of conservation activities could take place. These developments in museums were parallelled by the development of conservation management plans for monuments and sites.

The issues most concerning conservators at the beginning of the twenty-first century included the shaping of the conservation profession, the development of accreditation procedures and the need to work more closely with the 'public'. As views on heritage developed and changed, and as more people wished to be involved in different ways with the heritage, conservators had to balance the requirements of providing access *now* with the need to preserve cultural heritage for the *future*.

4 The meaning of objects

Introduction

This chapter is about the meanings and values attached to objects which provide the very reason for conservation.

> *The future challenges of the conservation field will stem not only from heritage objects and sites themselves but from the contexts in which society embeds them. These contexts, the values people draw from them, the functions heritage objects serve for society, the uses to which heritage is put, are the real source of the meaning of heritage, and the raison d'etre for conservation in all senses. (Avrami* et al *2000b, 4)*

Understanding the object prior to conservation is essential for both social and practical reasons. Things contain different meanings for different people or groups; they can also transmit different meanings in different contexts. At one end of the spectrum meanings may be concrete and widely understood, whereas at the other end they may be ambiguous, difficult to define or understood only by someone from a particular culture. It is important to understand the historical, aesthetic, religious or other context of an object before making any curatorial or conservation decision.

The perceived status of an object or monument will affect its importance within the material heritage of the times; for example it will affect whether it is acquired for a collection, or listed, or otherwise designated as a heritage object, and thus whether it is conserved. Most of the qualities which are taken as indicators of the object's importance are also qualities which can be damaged or distorted by inappropriate use or conservation treatments, so it is essential that these qualities are recognized and, as far as possible, understood.

Although this book focuses on the conservation of archaeological and museum objects, this chapter draws on work relating to buildings and art objects where there has been much consideration of values and meanings. Archaeological material, particularly in the hands of conservators, has been seen as having mainly evidence value but it is the intention of this discussion to look at the many other ways in which material culture can be valued, and so to prompt consideration of different possible meanings of archaeological and other 'non-art' objects.

Authenticity and other attributes

A range of terms is used to indicate the importance or meaning of objects; all are used in general discussion and debate, and all have been enshrined in charters or codes of practice relating to the cultural heritage. These terms are 'value', 'significance', 'integrity' and 'authenticity'.

Thus objects are considered to have different values or are described as having integrity or significance, or as being 'authentic'. Such ascribed qualities often play a crucial part in decisions about cultural heritage, and they may then affect conservation policies as well as practice. In this way, monuments and sites are accepted for national or international recognition and protection on the basis of their perceived cultural significance (the *Burra*

Charter) or of their presumed significance and authenticity (*World Heritage Convention*) or their ascribed significance and integrity (US *National Register of Historic Places*). Objects such as paintings and antiquities are investigated to find out whether they are structurally and materially authentic; conservators of objects are required by their codes of practice not to affect the integrity of the object on which they are working (National Historic Preservation Act 1966, UNESCO 1972, AIC 1996, Australia ICOMOS 1999, UKIC 1999).

While 'value' and 'significance' have been generally understood to cover a range of tangible and intangible meanings, 'authenticity' and 'integrity' were initially used to indicate originality and material completeness, but are now also used to embrace abstract attributes such as cultural or religious tradition.

Authenticity and integrity

Both authenticity and integrity encompass the concept of the original. Thus Jokilehto considers that 'authenticity cannot be added to the object; it can only be revealed in so far as it exists' (1995, 19), and Benjamin (1936) considers that authenticity requires the presence of the original. For many people, it is clear that the original has a particular power, 'the moral force of the real thing' (Pearce 1995, 154), but recent consideration of the concept of authenticity has questioned the claim that *only* the original has this power, or the ability to generate awe or fascination. For example the evaluation of authenticity used in selecting World Heritage sites has, until recently, been responsible for excluding monumental structures, recognized as highly significant in their own cultural context, built of perishable materials such as earth and wood which require regular renewal and replacement to survive (Larsen 1995). In this case the *Venice Charter*, which enjoins those responsible to preserve monuments 'in the full richness of their authenticity' (which has been taken to mean original), now appears oversimplified (ICOMOS 1964). Authenticity also encompasses function: for some artefacts such as clocks or industrial machinery, authenticity is demonstrated through the way they work, by telling the time, or spinning a thread. Maintaining working order may involve replacing original moving parts that are too worn to function.

There is a recent, but substantial interdisciplinary literature (referred to in Ucko 2000) which points to the new meanings being ascribed to 'copies' which maintain many of the qualities of the 'original'. In this context it is important to note that some museums do now acknowledge replicas (eg the Museum of the Diaspora in Tel Aviv, Israel: see Leigh 1994), as does heritage (eg the replica townscape of Warsaw: Jokilehto 1999, and see Ucko 2000). Thus it cannot any longer be simply accepted that authenticity alone provides 'that curious quality so critical to the mental and physical enjoyment and appreciation of historic works' (Matero 1993, 17). Some replication is faithful in all manufacturing details: for example, a twenty-first century piece of furniture which has been made in every way as if it was a sixteenth-century example using the appropriate type of wood, the same types of tools (which leave the expected tool-marks in the surface), and the same types of hand-made nail or dowel. Some 'authentic' artefacts, such as many of the existing, and flying, Spitfires (UK fighter aeroplanes from the Second World War) are in fact part-original and part-rebuilt, ie they are authentic replicas based on original blueprints (Holtorf & Schadla-Hall 1999). These are 'objects which depend for their appeal entirely on their ability to evoke some of the magic of the original' (Jones 1990, 29). It is important that conservation takes account of this new thinking about the power of replicas and the relative significance of the original or the 'authentic'.

Casts taken from famous works may become authentic in their own right, for example plaster casts made of classical sculptures, now accepted as important features of museum collections. Some of the plaster casts taken in the early nineteenth century from sculptures which are still in position on the Parthenon in Athens form a unique record of the condition of these sculptures two centuries ago (Plenderleith 1956). Copies may, of course, also play a role in conserving originals as when sculpture is removed from the exterior of a building to save it from further exposure to the weather, and a copy is put in its place – the copy then plays an important role in maintaining the authentic appearance of the exterior of the building.

Authenticity is complicated in situations where objects have been substantially modified by later restoration. Both the Warwick and the Piranesi vases were based on Roman fragments excavated from Hadrian's villa and restored in the late eighteenth century by Italian restorers or artists, including Piranesi (Marks & Blench 1979, Jones 1990, Miller 1992). Only parts are authentically Roman, so 'while it is perhaps inappropriate to describe the [Piranesi] vase as a forgery, its status as an antiquity must remain ambiguous' (Vaughan 1990, 133). However, it can be argued that these restorations became acclaimed and authentic art works in their own right (and the Warwick vase was much copied in the nineteenth century). Many buildings, of course, embody different phases, and later additions are not now seen to detract from the authenticity of the whole.

Authentication

Authenticity has been generally considered to mean genuine in terms of materials, workmanship and date, and processes used to authenticate objects concentrated on the identification of raw materials, the examination of tool-marks and other aspects of construction, and, where possible, the use of scientific dating techniques (Fleming 1975, Lambert 1997). Now the assessment of authenticity of an object or monument also takes account of the cultural context and both the material and conceptual expressions which are essential to the heritage of the particular group or society. This wider view is set out in the *Nara Document on Authenticity* (Larsen 1995).

When a thing is described as authentic this should indicate that it is what it purports to be, for example 'the product of an authentic tradition' or 'an authentic eighteenth-century cup'. Thus both material and conceptual aspects of authenticity are tested against a 'control group' of what is considered genuine for the time and the cultural context, just as a painting thought to be by a particular artist is compared to examples known to be by the same hand (Jones 1990, Phillips 1997). The process of establishing authenticity may enrich understanding of the past, but unless undertaken in a strictly research context a possible consequence is that it may be seen to enhance the economic value of the object concerned, and thus support the theft of, and illicit trade in, antiquities and works of art. Codes of practice such as that of the UK Institute for Conservation now clearly condemn the trade and consequently conservators and scientists are instructed not to undertake any action which might be seen to increase the commercial value of an object (UKIC 1999).

Significance

For conservators dealing not just with artefacts but with human remains, natural history and geological specimens, the concept of significance is often more relevant than that of authenticity. The significance of an object depends on the range of values or attributes which have been assigned to it at any one time, and it is subject to shifts of fashion and to change. When the term is used to describe an object it normally means its significance today. This may or may not be influenced by past or other significances, including original purpose, but it is today's significance which is normally used to discriminate between one object or place and another. Significance covers intangible as well as tangible attributes. An object or place can be described as having deeply spiritual significance for one group of people, or as having acquired significance because of certain events or associations, as in the case of the old city of Warsaw which acquired particular significance because of its deliberate destruction during the Second World War.

The significance of any object can be taken to be the different ways in which people interpret and use it. There may well be not just one single, homogenous, significance, since usually there will be several groups of people (stakeholders) who have an interest in the object. The originators, or their successors (eg indigenous peoples, Western artists, or a local community) may be concerned with the original meaning of the object and its 'appropriate' use or they may be concerned either with its current position within society, or even with how to re-incorporate the object within society. In other words, different groups within 'the public' may see the object as art, or as a means of making links with the 'past' or the 'other', or of creating a 'new' narrative. Owners of objects may see them as cultural resource, or purely decorative, or as functional (eg furniture), or as financial investment. Furthermore, there is another whole area of complexity, whereby the status of an object may be the subject of dispute, as in the looting and illicit trading of antiquities, or where requests have been made for the repatriation of objects which have a significance for the originators.

Clearly the concepts of significance and authenticity are linked but are not used interchangeably. Thus the significance of an object may not lie in its authenticity but in other qualities which provide information and enrich interpretations of the past. The views of stakeholders play an essential role in determining the aims of conservation treatment, so it is important that conservators are sensitive to the range of possible views about an object, and that they work with the stakeholders in reaching decisions about conservation.

Values

Objects and, particularly, monuments are often described as having different kinds of value such as 'age value', or 'art value'. The values which are discussed most frequently in relation to the material heritage include historic, artistic, scientific, cultural, contextual, condition and economic. Value has been discussed widely, and different authors have produced lists of values which they have used to identify the character of an object or building (Lipe 1984, Appadurai 1986, Pearce 1992, Matero 1993, Riegl 1996, Mohr & Schmidt 1997, Ashley-Smith 1999a, and see also Australia ICOMOS 1999 and Avrami *et al* 2000a). Much of this discussion has focused on buildings and monuments; another body of literature concerns the artistic or aesthetic value of works of art such as paintings or music (Budd 1995, Phillips 1997); and still another, derived from social anthropology, concerns the distribution of 'commodities' (Miller 1994).

In this sense, value is a qualitative and subjective assessment: one group may consider that an object has art value, but another group in another context might not see it as having anything to do with art.

> *In general value is a property that is assigned to an object in a manner that arises from the social context in question and it is to some, usually significant, extent arbitrary. It is never a property inherent within an object or material in the manner of such physical and measurable properties as hardness, density, refractive index, and so on. It cannot be measured outside a social context. (Renfrew 1986, 158)*

Value may change during an object's life. Some materials may have an initial value which they do not maintain, or start life with one value but acquire different values at a later stage. When a piece of clothing is acquired it has use value, this value will be lost when it becomes worn or no longer fashionable, but it may acquire sentimental value and subsequently rarity value and so become an 'heirloom'; the end of this stage of its life may be marked by its being collected by a museum.

Values are not mutually exclusive and an object can be seen to have more than one type of value, indeed Budd states that a work of art 'can possess as many kinds of value as there are points of view from which it can be evaluated' (Budd 1995, 1). Nevertheless, recognition of these values, of whatever kind, is important. They may frequently be the qualities which have led to an object being acquired by a museum or designated as being of national importance. They are the qualities which are exploited during use and which the conservator must attempt to ensure will not be lost (Fig 7).

Ascribing values is the approach set out in the *Burra Charter* for assessing the cultural significance of heritage places, and thus in aiding conservation decisions, and the same approach could be applied to objects and collections (Townley & Parris 1994). The concept of values can be useful to conservators, since once they are aware of the range of values ascribed to an object it becomes more possible to propose suitable conservation measures; however, whenever conservators examine and assess an object they must also be mindful that different cultures have different value systems.

Figure 7 Art or archaeology? Roman Mummy portrait which, as it is housed in an archaeological museum, and cared for by archaeological conservators, has not been fully restored as a portrait in an art gallery would be (Copyright: Petrie Museum of Egyptian Archaeology, University College London, UC 30088).

Economic value

In general parlance, value is frequently taken to mean economic value alone; however, until recently, heritage organizations claimed not normally to concern themselves with monetary value, except when they wished to acquire an item which would complete an aspect of a collection, or when a nationally important object was to be 'saved' for the nation by purchase. Most museum objects were considered to be outside the economic value sphere.

> *'Everyone' is against commoditizing what has been publicly marked as singular and made sacred: public parks, national landmarks, the Lincoln Memorial, George Washington's false teeth at Mount Vernon. (Kopytoff 1986, 77)*

Recently, however, the range of values ascribed to the heritage has shifted to include the concept of economic value. Policy groups and academic economists are attempting to put financial value on objects, monuments and collections, not only for insurance purposes but also to give substance to perceptions about 'value'. So, for example, in a study undertaken in 1998, financial criteria were used to assess whether to remove a road which impinged on Stonehenge, the prehistoric stone circle in the south of the UK. Members of the public were asked whether, and if so how much, they would be prepared to pay to reroute the road through a tunnel and out of sight of Stonehenge. Their willingness to pay was taken to indicate the value they put on Stonehenge itself (Maddison & Mourato 1999).

In 1999 the Courtauld Institute Galleries in London invited Sothebys, the auctioneers, to value some of the paintings on display and these values were added to the information on the gallery labels with the expressed intention that the relationship between art and price should be openly discussed (Lister 1999). Although this generated indignation from some who believed that 'art should not sully itself with such base considerations as prices' (Walter 1999), Ashley-Smith (1999a) has argued that financial value impinges in many ways on the conservation of museum collections, including the calculation of risk to the collections which might be caused by curatorial or conservation decisions. Conservators, increasingly working independently, have to price their own work, and fees charged for conservation will inevitably have some relation to the economic value of the objects conserved.

Classification and labelling

The terminology used to classify or describe things attaches a meaning which almost certainly will influence attitudes. Whereas terms such as 'thing' and 'object' may be fairly neutral, 'artefact' and 'structure' imply manufacture, while 'goods' and 'commodities' usually have economic value, and 'monument' and 'work of art' may be intended to impart a message or to have some form of visual impact (Appadurai 1986, Kopytoff 1986, Pearce 1992). Such labels may change, or be interpreted differently by different people. An ecofact, once found, will become a specimen, be identified and put into a collection; an ethnographic artefact in one collection may become 'primitive art' in another, a painting hitherto labelled 'school of' may be acclaimed as the work of the artist him/herself after expert reassessment; a building may become a monument as a result of events which take place in it (Phillips 1997, Jokilehto 1999). Labels can be powerful indicators of significance, and so may prevent the viewer (including the conservator) from making an unfettered appraisal. They can also raise questions, as when heritage objects become commodities and are subject to looting and sale (Kopytoff 1986), or when human remains housed in museums are regarded as specimens, to the great distress of the descendants:

> *As we crossed the country and visited the universities, museums and laboratories, we found the bodies of our ancestors stored in cardboard boxes, plastic bags and paper sacks. We*

found our sacred burial places stripped and desecrated, the bodies and sacred objects buried with our dead on display for the curious and labelled 'collections', 'specimens' and 'objects of antiquities'. (Hammil & Cruz 1989, 195)

Apart from understanding the 'power' of labels, it is also important to realize that it is not necessarily possible or desirable to classify every object, that there are often extra 'hidden' meanings, and there are certainly cultural differences in attitudes to understanding objects.

Lives of objects

Objects are often described as having lives. An object's life starts at the time of formation or manufacture and then passes through a varying number of stages and transformations, and eventually may end in disintegration. Pearce described this in the following passage.

As it moves through time, it acquires (to a greater or lesser extent) a history of its own, passing from one possessor to another, perhaps from one kind of use to another, from one place to another. Some objects, especially those to which special values are attributed, have very long and complex life cycles. Other objects, especially some of the mass production material of our time (like packaging) are discarded immediately, to be taken away in the next rubbish collection and buried or burned at the corporation dump. Objects can have complicated life cycles in which part of their time is passed in enjoying some kind of regard, part as garbage or semi-garbage in discard dumps or junk heaps and part resurrected by archaeological processes or by shifts of taste, to be reinstated in museum collections and the like. (Pearce 1992, 16)

Of interest here are the changes in status or importance to which an object may be subjected during its 'life-time', some of which are the effects of use, some of fashion and so on. These changes are often associated with a change in the perception of the value of the object so, for example, an ordinary utilitarian object may acquire sentimental value because of its associations with a person or an event, or the mass-produced packaging described by Pearce as the kind of material which is normally considered to be rubbish may be made the subject of museum display and so become collectable. It seems clear that even apparently major monuments such as Stonehenge may be considered 'important' at one time and 'not worth bothering about' at others (Ucko 1989, xiii). The change of status when an object is taken into a museum, and shifts which may occur within the museum as perceptions and use of objects or collections change, are particularly important to curator and conservator.

The life of an African mask in its original cultural context, and subsequently in a museum, shows how many material and conceptual changes an object may pass through:

It, the mask and the being it shows, or embodies, may appear periodically. For some of its life the mask will be stored away, possibly hidden, then it will be brought out and danced. At each stage of its life its appearance will be different. Before being worn, it may be repainted, mended and have magical substances added to it, have sacrifices made over it; for the actual dancing it will have a skirt or costume attached to it. Sometimes masquerade costumes are made for a single use, a matter of a few hours only, made of the most ephemeral materials such as freshly picked leaves. After the ritual the mask may be kept, the skirt discarded. Over time the mask itself will become worn, encrusted with paint, be mended, have bits added or knocked off. (McLeod 1995, 7)

McLeod goes on to remind his audience that the custom in European museums until into the second half of the twentieth century would have been to remove the accumulated sacrificial deposits in order to expose the original carving (McLeod 1995), presumably treating the mask as 'art'. This would have shown little regard for the conceptual integrity of the object and would have removed much of the material evidence which demonstrated the significance of the mask, and the history of its use.

Thus the details of the life of the object may contribute to the nature of its changing significance within a society and so to the nature of its preservation. The vagaries of the life of the object result in the accumulated evidence of manufacture, use and alteration, some or all of which may make the object of interest and worth preserving, and this is the evidence which needs to be understood as much by the conservator as by the archaeologist, historian or other specialist (Fig 8). Kopytoff (1986) discussed the idea of objects having biographies. The use of the term 'biography' conjures up very effectively the changes and chances of an object's life and the great variety of effects on the object itself. In the case of archaeological and museum objects the changes of significance which an object passes through could be called its 'conceptual biography', and the changes in physical condition could be its 'material biography'.

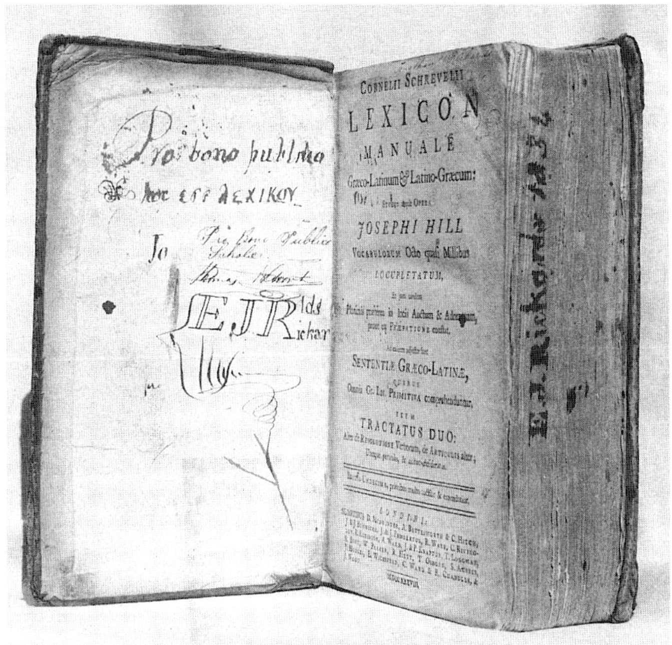

Figure 8 History or damage? Greek Lexicon published in 1738 and covered with schoolboy scribbles in 1834. Is this damage or is this history? Doodles done today would be seen as damage but these scribbles are now an important part of the significance of the object (Courtesy of N D Balaam).

Different ways of understanding objects

Is there a single truth about an object?

In the twenty-first century conservators (and curators and archaeologists) need to adopt as a working premise that there is no such thing as a single truth about an object. Firstly, people look for, and see, different meanings in the same object, and, secondly, each object has undergone so many 'life changes' and functional modifications that its 'life history' must be elucidated, and not just assumed, by materials scientists or conservators. It is dangerous to think that as long as conservators proceed cautiously and watch carefully the, or a, 'truth' will become apparent beneath the swab or the scalpel. Information about an object revealed during examination and conservation requires interpretation before it can be properly described and recorded. New information can lead to the significance of an object being re-evaluated or reinterpreted, and in turn this can lead to a shift in the object's role or position within any reconstruction of the past.

> *Things are equivocal. A pot can be classified according to its shape and decoration as of a particular type. But thin-sectioned under a polarising microscope it explodes into another world of micro-particles and mineral inclusions. The pot is not just one thing which can be captured in a single all-encompassing definition. There is always something more that can be said or done with the pot. A single pot is also multiple. It depends on the trials we make of it, what we do with it, how we experience it, whether we attend to surface or shape, or slice it and magnify it. (Shanks & Hodder 1995, 9)*

None of the terms used to describe objects is a constant, in that views about objects, and knowledge of their structure and function, change and develop through time, with fashion and with the development of new investigative techniques.

Material significance

Questions of significance, integrity or authenticity tend to be viewed by archaeological conservators and materials scientists from the point of view of the structure and composition of an object. Their interests traditionally centre on the evidence of raw materials, manufacturing processes and modifications embedded in the object itself, rather than on any conceptual or cultural aspects.

Taking ethnographic objects as an example, their material significance depends, in part, on the original manufacture and, in part, on the effects of collection and housing in a museum. Apart from traded materials (which might include metals or glass beads), the raw materials used in manufacturing the objects often depended on the locally available plant and animal life. Artefacts from very cold climates where there were few plants and no trees were made from animal materials (see Morrison 1986). Conversely, objects from tropical areas where trees and plants were abundant were normally made of plant materials (Florian 1990). Once they entered museums, the objects were often seen as replaceable, so compared, say, to European or Oriental works of art they tended to be stored in poor conditions. Many were treated at some point in the past with insecticides but there may be no records of what was used and the toxic insecticides remain in the objects (Johnson 1999). Both raw materials and technologies may be unfamiliar, and their reactions to the environment of a centrally-heated museum may add further difficulties. Thus, simply from the material point of view, ethnographic objects pose a range of problems to conservators.

Conservation processes inevitably affect the material integrity of any object, particularly if they involve addition of conservation substances or removal of accretions, hence the emphasis on preventive conservation and minimizing intervention in several charters and codes of practice (eg UKIC 1999). Removal of materials is normally undertaken to enhance the understanding and appreciation of the object by revealing something of its original material form, but this process is seldom straightforward. Removal of corrosion products on an excavated object always provides a dilemma: it may not be possible to understand the significance of the object without freeing it from obscuring layers of corrosion but the corrosion is materially part of the original, is evidence of its burial, and may contain traces of associated structures (Scott D A, 1994).

On the other hand, renewal or replacement of original materials when they are so degraded as to be no longer able to perform their role in the object has been an acceptable practice in situations where the continuing function of the object is considered to take precedence over material integrity. This situation includes replacement of bindings on books, and of working parts in industrial machinery (and here there are even arguments for maintaining working capacity by 'cannibalizing' identical machinery; see Ellam 1999). In some cultures material renewal of monuments is not seen as affecting significance. This is behind the dispute about the authenticity of wooden temple buildings in Japan or earthen buildings in Africa, which need regular renewal in the face of the onslaught of decay and disintegration (von Droste & Bertilsson 1995, Ito 1995).

Material evidence revealed (or obliterated) during a conservation process can be crucial. The debate about the Rubens painting *Samson and Delilah*, now in the National Gallery, London, revolves around the claim that at some stage when the painting was put on a new support (a normal restoration procedure when the original support is weakened) a panel-maker's stamp, which would have proved the painting to be a copy of a later date, may have been removed (see Mendick 2000). Most critics, however, are convinced of the attribution to Rubens (Sewell 2000).

All conservators are inevitably concerned with the material, even in situations where the intangible nature of the object (eg the aesthetic character of a painting) has long been considered crucial.

> *Through their daily work, most conservators are used to shifting constantly between the aesthetic appreciation of a work of art and the consideration of its material aspects. This switch is often harder for the art historian, schooled perhaps in what might be described as the 'otherness' of art. (Leighton 1998, 31)*

Evidence of history

Almost all significant heritage objects and places have been changed by the effects of time and use, if not by early repairs and modifications. Riegl argued that 'age-value' is often at variance with historical value (by which he meant the original state of the monument or object) (Riegl 1996). Thus authenticity might be thought to be compromised when an object has deteriorated or been altered in some other way.

The problem of change can be particularly acute when considering buildings and monuments which are inevitably exposed to the weather. Most early examples which remain today are either severely altered by deterioration (and may even be ruinous), or have been altered by being maintained and repaired during their lives in order to keep them standing and functioning. Many of the maintenance techniques used now or in the past have affected material integrity, but have added to the evidence of the history of the structure.

The evidence of history could be said to be 'written' in the materials of the object. Considering objects to be documents which provide information about technology, use, modification and deterioration is a familiar concept in archaeological conservation, and this is echoed by Petzet who sees monuments as 'an archive of authentic sources of cultural history, social history, industrial history, etc' (Petzet 1995, 87). It is acknowledged now that deliberate modifications as well as the effects of age and deterioration are an essential aspect of significance (and authenticity).

> *Now a new concept of authenticity is emerging which encourages us to accept that objects have a continuing history, that they are damaged and repaired, cleaned and restored, and that their present state records not only the moment of creation but also a whole subsequent sequence of events. (Jones 1990, 14)*

For conservators the challenge is to be alert to the indicators of such a life-history and to retain them, as far as it is possible to do so, without allowing further deterioration. It is also important that visible structural repairs or replacements are finished in such a way that they are sympathetic to the signs of age (Bristow 1990).

Conceptual integrity

Although it is well established that a work of art is 'more than' simply an assemblage of different materials (eg more than paint, varnish and canvas) the intangible qualities of other types of object have not been so widely recognized. The discussion of authenticity, significance and values indicates that objects are often perceived to have additional layers of meaning depending on their cultural context and related, say, to their role in national or personal history, or their sacred or aesthetic characteristics (McBryde 1985, Greene 1992, Reedy 1992). It is now increasingly acknowledged that the significance of a wide range of places and objects lies in their intangible or conceptual attributes rather than their concrete form. This is usually known as conceptual integrity or conceptual significance.

Clavir has defined conceptual integrity as encompassing:

> *the non-material properties of objects, properties such as religious or cultural significance, or the intention of the artist. These properties are included even if they are not physically evident to us through the object. (Clavir 1994, 53)*

Acknowledging the importance of the conceptual or cultural significance is relatively recent in the West (McCredie 1991). The Canadian Group of IIC added the importance of respecting conceptual integrity to their code of ethics in 1989 (IIC-CG 1989), and ICOMOS New Zealand issued the *Charter for the Conservation of Places of Cultural Heritage Value* in 1992, which speaks of indigenous cultural heritage as being 'inseparable from identity and well-being' and having 'particular cultural meanings' (ICOMOS New Zealand 1992). It was not until 1994 that an attempt was made to establish intangible cultural significance as a valid basis for recognizing World Heritage sites. This resulted in the *Nara Document on Authenticity* which states that 'all cultures and societies are rooted in the particular forms and means of tangibles and intangible expression which constitute their heritage, and these should be respected', and 'cultural heritage must be considered and judged within the cultural contexts to which it belongs' (Larsen 1995, xxii, xxiii). The *Document* acknowledged the fact that heavy reliance on material authenticity excluded monuments of major significance in their own culture which are regularly renewed or rebuilt.

Of particular interest are the changing attitudes to ethnographic objects. In most cases ethnographic objects were collected in order to study or exemplify 'the other' or a 'strange culture'. This unfamiliarity resulted in shifts of meaning attached to an object from 'curiosity', to 'primitive art' to 'ethnographic artefact', but now that indigenous groups are actively reviving their languages and customs they are also reclaiming the original meanings and roles of their own material culture, and sometimes seeking to reclaim the objects themselves (Faber 1993, Simpson 1996, Errington 1998). The conceptual integrity of many ethnographic objects derives from their original function in secret or sacred activities, which in turn affected who would have had access to them, and how they would have been handled and cared for in the original culture.

Most indigenous peoples wish to see their culture treated with greater respect, many want to have greater access to their objects in museums and to play a role in the way they are used and cared for. Some objects are repatriated, eg in North America if they were associated with a burial and so affected by the *Native American Graves Protection Act* (NAGPRA 1990). For conservators the issues include how to respond to requests that objects be borrowed for ceremonies, whether it is possible to segregate objects or store them in particular ways (Drumheller & Kaminitz 1994), whether certain objects should not be handled by specific people (eg women, or the uninitiated) (Idiens 1995) and whether some objects which were not intended to last should be given no interventive conservation treatment. There is also the need to ensure that nothing done to an object will prevent its resuming its spiritual role should it be borrowed by, or repatriated, to the originating group (Wolfe & Mibach 1983, Mellor 1992, McLeod 1995).

In areas where museum collections contain material relating to local indigenous peoples some fruitful collaborative arrangements have been established between communities and museums. In Australia conservators are advising aboriginal peoples about the care of their objects now transferred to local Keeping Places (Coote 1998). In Canada and North America collaboration can involve shared custodianship, access to objects in order to revive traditional crafts and skills, and loan of objects for use in ceremonies (Wolfe & Mibach 1983, Simpson 1996, Paterson & Greenfield 1998, Peers 1999). Where museums, such as the British Museum, hold collections from other parts of the world there is increasing consultation and involvement of interest groups when planning the objects' use in a new exhibition (eg Starzecka 1996, King 1999).

Artist's intent and attribution

Bound up with authenticity of art objects is the quality known as artist's intent (Dykstra 1996, Richardson 1996, Phillips 1997). Almost certainly with old objects this intent has been distorted not just by age but by repairs and restorations, and it is impossible to 'know' what a past artist intended unless she or he left detailed descriptions (Hedley 1990). Paintings also present particular difficulties in terms of identifying the artist, and the business of attributing them to specific artists is subjective, shifting and often controversial. A good example is provided by the situation with paintings by Rembrandt. A research project was initiated in the late 1960s in the Netherlands 'because of the exceptionally contentious state of Rembrandt attributions. In nearly a century of successive *catalogues raisonnés*, the paintings attributed to him had risen from 595 to 711, then declined again to 430' (Phillips 1997, 77).

Copies and authenticity

Attribution of paintings to specific artists is complicated by the fact that copying has been an accepted and widespread art practice. Thus several versions of, say, an eighteenth-century portrait of a prominent person, may exist by the same artist or by his/her assistants.

> *in many cultures copying has often been the dominant mode of artistic activity, motivated by a desire to maintain or renew traditional forms and skills, by nostalgia for the past and admiration of its achievements. (Jones 1990, 29)*

Some art works such as engravings and etchings (and some sculptures) are, by their very nature, produced in multiples (as are industrially produced objects, for example many of those in the collections of the Victoria and Albert Museum, London). Each one can be said to be authentic, and the debate shifts to which was the earliest. Complex situations can arise as in the case of the bronze casts made after Degas' death from his original wax sculptures, including *Little Dancer Aged Fourteen*, bronze versions of which now exist in many major museum collections (Rolfe 1999). Many copies were not meant to deceive, in fact many were intended to celebrate the original.

Artist's intent: contemporary and conceptual art

Just as objects from a 'strange culture' are a challenge to a conservator, so too may be items of contemporary art, some of which present new challenges to the concept of integrity or authenticity, and to the way in which these objects are cared for and displayed in museums or public collections (Albano 1996). Much contemporary art is experimental in the sense of using new materials or technologies, such as Gabo's use of plastics in his sculpture during the 1930s (Pullen 1999). Even more challenging is the work of artists who have explored concepts such as change and decay through the deliberate incorporation of highly perishable materials such as chocolate or other foodstuffs (Magor 1990, Aben 1995, Wharton *et al* 1995, Heuman 1999a). Some artists use their art to convey other types of message, such as *Pharmacopoeia* by Freeman and Lee which incorporates brightly coloured pills and capsules to highlight the issue of social dependence on medical drugs (Collier 2000).

Conceptual, particularly auto-destructive, art presents the conservator with the challenge of preserving the concept, and questions which arise include to what extent the artist may be involved, indeed whether maintenance and conservation become part and parcel of the artistic process itself (Magor 1990). Given the problems of rapid deterioration of many of the materials used, the conservator may have to decide whether to replace materials or simply let them decay, effectively letting the object die. A further question arises if the object is appreciably changed through replacement of components, or through reworking by the artist: does it remain the same object, or should it be redated? (Smith 1990, Heuman 1999b). Not only do the concepts and the materials provide a challenge but the primacy of the artist's intent can bring with it questions of the artist's copyright and his/her legal rights over the object and its conservation, even if it has been purchased by a museum (Curnoe 1990, Nemiroff 1990, Stone 1990, Fry 1997).

Authenticity of performance

Authenticity of skill or of performance may be linked to the preservation of the integrity or authenticity of an object or site. Audiences may have very different expectations about what constitutes an acceptable 'performance' (Ucko 2000), but it is generally agreed that the ability to demonstrate the process is an important aspect of the authenticity of industrial machinery.

> *if industrial collections and museums are to preserve processes as well as artefacts, they must do more than maintain them in glass cases. In contrast to collections of paintings or sculptures, the artefacts have rarely been collected for their aesthetic value alone.* (Ellam 1999, 4)

Industrial machinery needed the care of technicians or engineers to keep it functioning effectively during its working life, and these same skills (rather than conservation skills) are needed to keep machines in working order and to operate them for an audience: without these skills they may simply not function at all. However, keeping a machine or a clock running may mean that some of the original moving parts become so worn that they have to be replaced, thus the material integrity is impaired while the machine's ability to demonstrate the authentic process is maintained (Brodie 1994, Keene 1994, Mann 1994a, Mann 1994b). For these reasons there is considerable discussion of whether historic musical instruments should be maintained in playable condition (Barclay 1997).

Authentic skills and technologies (such as thatching) are also needed to maintain the integrity of traditional buildings (see Hawkins & Macdonald 1997). In some cases both industrial and building skills are dying out as the last 'genuine' practitioners reach old age, and so may need to be conserved or revived, eg through training courses for builders, architects and volunteers.

Contested material

Museums also contain some contested materials which, however, until relatively recently, were not generally considered to be problematic. These are human remains, and looted or illegally traded objects. Neither presents a new problem but each has become a major issue through the work of international, national or local pressure groups. Both human remains and looted and illicitly traded objects are the subject of requests for return (repatriation) to their owners and many or most museums are now accepting that they must treat such requests sympathetically. The *Native American Graves Protection Act* (NAGPRA) passed in 1990 provides a good example of a national approach to dealing with sensitive materials housed in museums, as does the range of measures in Europe designed to identify and return art works looted during the Second World War.

Human remains

Although human remains do not normally present problems of material authenticity, they may be highly significant and their presence in museums or study collections has become an emotive issue. In the past human remains were acquired as curiosities (human mummies) or as specimens for scientific study, particularly in medical and natural history collections (Fforde 1992, Ucko 1992).

Museums may also hold relatively recently excavated material (often from early cemeteries exposed during modern building work); for example the Museum of London was known to be holding 6500 skeletons in 1998 (White & Ganiaris 1998).

Until comparatively recently many of these remains were treated with little respect or any consideration for what might have been the wishes of the people themselves (Layton 1989b, Simpson 1996). This had become the situation even though in most societies it is considered desecration to disturb a grave, and when, for some cultures, keeping human remains unburied represents a major disruption of the journey to the after-life:

> *Any disruption, delay or halt in that journey is a violation of personal religious beliefs to that individual, to his descendants who incorporate and are responsible for his spirit in their daily lives and religious ceremonies, and to those of the present and the future who will embark on that journey. (Hammil & Cruz 1989, 195–6)*

Now, at the beginning of the twenty-first century, the problem lies in trying to balance the desire for genuine and productive scientific study with the sensitive nature of the remains. Scientific study can contribute to understanding of past populations, social factors such as diet and living conditions, and medical questions such as origins and spread of disease, all of which may be of interest not only to archaeologists or anthropologists but also to the local community. Although the ICOM Code of Practice provides guidance on the museum care of human remains, this problem was not fully addressed until the *Vermillion Accord* was drawn up in 1989. This accord was one of the results of the World Archaeological Congress' First Inter-Congress held at Vermillion, South Dakota, USA, and it attempts to reconcile the conflicting issues of proper respect for the dead and study of their remains. Its final paragraph states that 'concerns of various ethnic groups, as well as those of science are legitimate and to be respected' (Hubert 1989a, Hubert 1989b).

The fact that there are large numbers of human remains in museum stores, and in some displays, means that conservators are inevitably responsible for their care. It has been normal to approach skeletal materials with a dispassionate and scientific attitude and to concentrate on understanding of bone as a material, as well as of the aims of scientific study. Attitudes are already changing in many museums and it is important that conservators also understand the cultural significance of the remains and devise methods of handling and storage which maintain respect for their status as *human* remains.

Looted or illegally traded material

Robbery and looting of archaeological sites, religious buildings, private collections and even of museums is not new but the problem has increased with the increasing commercial value of art and antiquities. In some parts of the world, eg Mali in West Africa, the cultural heritage has been stripped from sites and museums (Brent 1996). Nowhere in the world are archaeological sites entirely safe from clandestine digging and removal of antiquities, and wherever there is armed conflict looting of private and public buildings is inevitable. The trade is encouraged by the lucrative market for such objects and by the fact that private collectors and some museums have been willing to buy the material without attempting to check where it has come from (Fforde 1991, Tubb 1995, Simpson 1996, Vitelli 1996, Brodie *et al* 2000, McManamon & Morton 2000).

Conservators may be asked to work on recently stolen or looted objects, and since the intention of conservation is to elucidate objects which are obscured in some way, and to maintain or improve their condition, these actions enhance the monetary value of the object and thus aid the whole process of looting and trade. It is important that conservators' initial assessment of an object should include checking, and, if necessary, questioning, details of provenance and ownership. Any action which can be construed as providing a date or some other form of authenticity will also enhance the value, this can include an 'innocuous' condition report which identifies the type of object or describes the deterioration products as what might be expected of a certain context. Unfortunately it is also clear that unscrupulous conservators are involved in the looting and subsequent preparation of material for sale. This is apparent from artefacts such as mosaics which would have needed specialist preparation before theft if they were to survive at all (Tubb 1995, Jaeschke 1996, Tubb & Sease 1996, Sease 1997).

Attempts have been made at an international level to stop this trade, eg through the UNESCO *Convention* adopted in 1970, and the UNIDROIT *Convention* adopted in 1995, but unfortunately there are still countries (including the UK) which have not adopted these conventions. Many professional organizations include clauses in their codes of practice condemning this trade (eg ICOM) or enjoining members not to work on looted material (eg UKIC). There is greater awareness of the iniquities of the trade and, in museums, greater caution in checking objects before considering purchase. Some museums have been shamed into admitting possession of looted material, and improved documentation of cultural heritage is making it more difficult to conceal the origin of stolen material (Thornes 1995, Thornes 1997). Sadly, archaeological material cannot be documented while it remains undiscovered.

Objects and context

Context has an effect on significance or authenticity of an object or structure.

> *Context refers to an object's immediate surroundings, inasmuch as these determine the approach and, thus, the correct interpretation of the object; that is, the frame of a picture, traditional surroundings of a monument that are essential to its scale and significance, and social circumstances in which the object is or was used. (Philippot 1976, 370–1)*

Archaeological objects derive much of their significance from their relationship to the context (structure of site, other objects) in which they are found. Some artefacts are designed and made for a specific context, such as wall paintings, window glass or some furniture, and for some the meaning and function is related to context, such as a lighthouse built on top of a rocky shoreline. Some gain impact from location, eg a major building and piazza set in a network of medieval streets, or a religious painting hung in a church. Context 'provides scale, mass, texture, volumetric form and style' (Warren 1998, 14) – to these could also be added particular effects of light and colour. Buildings gain authenticity from their contents, eg furniture and fittings. Any change in these interrelationships can affect significance.

Removal from the original context affects authenticity even if the object itself is more or less unchanged (Fig 9). Looted or stolen objects are forcibly removed from their proper context and associated material, and are presented to a purchaser with no information on provenance (Sease 1997). With archaeological objects, even if the context is fully recorded, 'the evidence has largely lost its original integrity. Its meaning rests only in the information attached to it in the form of associated records' (Crowther 1989, 42). Excavated objects are in danger of deteriorating after removal from their burial context, thus further affecting their integrity. Other objects, such as the Maori house now in the garden of Clandon Park in Surrey, UK, a National Trust historic house (Hooper Greenhill 1998), or wall paintings removed and displayed in a museum like easel paintings, may have lost much of their original meaning. While the wall painting might survive better materially after being detached from a damp or disintegrating wall, it will certainly have lost much of its conceptual significance, and the Maori house deteriorates conceptually if

Figure 9 Loss of context: Temple Bar, a seventeenth-century gateway to the City of London reputedly designed by Sir Christopher Wren. It was removed in the late nineteenth century to make way for traffic and now stands isolated in a park in Hertfordshire, UK. There are proposals to return it to the City of London (although not to its original position) (Courtesy of N D Balaam).

it is not used in traditional ways by a Maori community.

Modification of the context can be almost as undermining of authenticity as removal, as in the case of a single remaining medieval building now surrounded by modern office-blocks, or an over tidy, much-labelled and charmless historic site (Lowenthal 1979, Morin 1999). Attempts to mark the significance of an object or building through modifying the context may be seen by some, at least, to have had the opposite effect – the log-cabin in which Abraham Lincoln is said to have been born seems overwhelmed by the monumental, temple-like stone building erected round it (Fitch 1982).

Context is maintained by preservation of landscape and 'ordinary' buildings together with major monuments. Furniture and fittings can be returned to the intended context, as in the refurnishing of Chiswick House (an early nineteenth-century Palladian-style villa on the outskirts of London owned by English Heritage (Drysdale & Rowe 1998). Museums can attempt to mimic a context through room settings, or reconstructions of village or industrial complexes in open-air museums, but these are inevitably artificial (Kirshenblatt-Gimblett 1991). For conservators the challenge is to attempt to preserve objects in their contexts; to devise methods for keeping artefacts such as wall paintings and window glass *in situ* and to treat assemblages and collections as unities (Wright & Player-Dahnsjö 1998, Seeley 1999).

Context: what happens when an object is placed in a museum?

'Is your museum a mausoleum?' was the question posed by Hugh Casson when addressing the UK Museums Association Conference in 1960, nearly half a century ago (Strong 1999). Some do see museums as full of dead objects, and there is general agreement that objects change, or are changed, when they enter a museum. If they do not die, their 'normal' lives are certainly interrupted, and they are experienced differently from before.

> *When museums and art galleries acquire contemporary works of art, the works very often go into a state of suspended animation due to the parameters applied to all works of art*

that enter the institution. The very word 'museum' suggests the beginning of the end. (Barclay 1990, 28)

These protective measures alter the conditions in which artefacts are experienced: they remove relics from the here and now, from continuity with the world around them, to an exclusive milieu. (Lowenthal 1979, 112)

Comparing natural heritage with cultural heritage Cramer points out that the aim of preservation of the natural heritage should be to 'restore the evolutionary adaptability of nature, the capacity to rejuvenate itself' while 'even the best museum can only present objects taken out of context, in location, in time and in culture. In this respect, a museum takes on the character of a tomb' (Cramer 1994, 23). In museums many dynamic objects become stilled (Figs 10 and 11), such as costume and personal ornaments which were intended to be seen and appreciated, at least in part, in movement; or clocks which no longer function. Objects are 'withdrawn from their exchange sphere and deactivated, so to speak, as commodities' (Kopytoff 1986, 76; and see Pearce 1992).

Figures 10 and 11 Museum object or ceremonial object? The eighteenth-century Lord Mayor's coach (Corporation of London) spends most of its life as a museum object in the Museum of London, but each year it is used for the Lord Mayor's Show where its has to function as it always did, drawn by horses through the streets of London. It has retained its significance as a ceremonial object (Courtesy of Museum of London).

This effect, of life and continuity being interrupted, is felt acutely by peoples who object to their ancestral human remains being removed to museums, and to their secret or sacred objects being handled or exposed to view in inappropriate ways (Hammil & Cruz 1989), and by contemporary artists who see their art as evolving rather than static.

Conversely, the act of collecting an object can be seen as saving it and extending its life, perhaps beyond what would normally be possible. 'There is also something of the act of saving, of some sort of life which would otherwise not be. The collected things are "saved". They would otherwise be dead' (Shanks 1992, 59). Objects can be seen as validated by being acquired by a museum: 'they have properties bestowed on them by virtue of their museum existence' (Gathercole 1989, 74). Illicitly traded material acquires apparent respectability, ephemera become museum objects and rubbish becomes durable (Kopytoff 1986, Pearce 1992, Ashley-Smith 1999a). They 'share a perceived spiritual or intellectual worth' (Pearce 1992, 33) and are 'scrutinised and cherished, sitting in a vortex of academic and popular attention, safe behind the glass of a museum display case' (Crowther 1989, 43).

The way objects are displayed in a museum can influence their perceived significance; for example an African mask may be displayed and interpreted in terms of its meaning and use within the culture from which it came (ie as an ethnographic object), or it may be displayed with the minimum of explanation, as an art object expected to 'speak for itself' (Leyten 1993, Simpson 1996, Errington 1998).

Not everything in a museum *is* on display, so another effect of museum acquisition can be to make some objects, apparently or actually inaccessible 'locked away in store-rooms, revealing their secrets only to the initiated' (Gathercole 1989, 76). Archaeological archives, the collected records, environmental samples and artefacts from excavations, often now housed in museums, represent a large group of material which is effectively 'frozen' and unused, though part may be reactivated and become a useful resource when a research project focuses on it (Carver 1996). Just as many museums are making their stores accessible to non-specialists, that is 'the public', so are museum archaeologists discussing ways of making the archaeological archive more widely available to universities and to schools (Merriman & Swain 1999).

By virtue of entering a museum, objects take on a new life. They are exposed to a range of new experiences such as the threat of what conservators and curators term 'museum dirt' (most major museums are in highly polluted cities though many now have air-conditioning), or of investigative procedures. They are used to serve the museum's political ends in the way that they are displayed, and major objects have restless lives in that they are more displayed, more lent and more restored (and investigated) than less significant ones. A recent paper on Rembrandt's painting *Anatomy Lesson of Dr Nicholaes Tulp*, painted in 1632 and always in an institutional or Royal collection, revealed that it had been treated at least 23 times since 1700 (Wadum & Noble 1999).

Objects are considered to be inalienable (so cannot normally be transferred to a different purpose or ownership) while in a museum. Some, however, may be de-accessioned and leave the museum either because they are no longer relevant to that museum or no longer usable (through decay), or because they are returned though repatriation to their originators (and so returned to their original significance and to the normal course of their life which may mean to reburial, or original use, eg in ceremony).

Meanings and conservation

How is this relevant to conservation? Conservators should put more emphasis on establishing the range of possible values which objects have for different groups, and they should be familiar with all the possible influences on significance. Artefacts which have hitherto been seen as having purely material or evidential significance (such as many archaeological objects) should be considered in the light of the layers of meaning which are attached to other types of object.

It is clear that the complexity of attitudes and ascribed values to what the West has termed 'authentic' or 'copy', 'replica' or 'genuine', are being better understood than ever before. Changes in understanding of such concepts, being explored and developed by lawyers, anthropologists and archaeologists with regard to copyright matters, will force reappraisal of assumptions held by conservators. The deconstruction of so many concepts and procedures, until now dominated by eighteenth- and nineteenth-century preconceptions, will also challenge many of the assumptions which form the core of conservation attitudes and practice.

5 Change in materials and objects

Introduction

This chapter is about the material changes which occur during the manufacture, use and deterioration of objects and it introduces the concept of conservation as the management of change. In the absence of human intervention there is a natural progression of change in all materials, from formation through maturity, to decay and ultimately disintegration. Thus change is a characteristic of the life of all artefacts, and many forms of change are inevitable.

Conservators need to understand and, as far as possible, recognize all the changes an object has passed through in order to assess its state and decide on conservation procedures. Processes of manufacture and effects of use and of history may add meaning to an object, but some deterioration may detract from it. Tracing changes in structure and composition could be likened to recording the 'material biography' of the object, and conservators must understand the role of material change in both gain and loss of significance.

Nature of change

Some effects of change are minor and imperceptible, some are so fundamental that the original material (or object) is barely recognizable. Some changes are intentional and controlled, as in the processes of manufacture; some are unintended and largely uncontrolled, as in deterioration and disintegration. Some forms of deterioration are accepted, or even welcomed (at a later date), for example the 'pleasing decay'of a semi-ruined monument, or a beautiful patina on the surface of a metal object. Some changes are considered part of the character and significance of the object, as in some contemporary art objects which are intended to decay and are planned to have parts replaced.

Raw materials are initially changed by extraction or removal from their source and by the processing needed to make them suitable for conversion into usable objects. The converted material, be it metal, skin, clay or whatever else, may then be changed again when it is formed into an artefact. It may be worked with tools, combined with other materials, shaped and decorated. Once created, an artefact is changed by use, wear, modification, maintenance and repair; discarded or abandoned artefacts are further changed by erosion, breakage, crushing and burial. Clearance or excavation of a site, or acquisition by a museum, bring about more changes by exposing the artefact to new and different conditions such as higher light levels, changes of temperature, or pollutants. Past attempts at restoration and conservation may have brought about short-term changes which might have been intended as improvements to the condition or legibility but may have set in train a series of later and less desirable changes. Modern conservation may change the object yet again, and there will undoubtedly be further changes in the future.

The significance and use of an object can also be affected by the various changes which have taken place, for example poor condition may mask significance. Potential for use can be modified or enhanced when the process of investigation, prior to conservation, reveals information which allows the object to be more fully understood, or reinterpreted. The way materials have been modified and used in the manufacture of an object may be of particular significance and may be appreciated only as a result of detailed investigation. This is a common situation with archaeological objects which may be virtually indecipherable, and so unusable for display or interpretation, prior to conservation (Keene 1987, Cronyn 1990) but it can also occur with other types of objects such as buildings, or paintings. For instance, investigation of the structure and condition of a painted surface may reveal the existence of earlier painting beneath, which changes the interpretation, and the use, of the whole (McClure 1992).

Given the inevitability of change, the conservator's task can be described as management of change. However, conservation should not obliterate the beauty and fascination of change, nor the information to be derived from it, and it should not 'preserve in aspic' and prevent changes which reflect current processes of social or environmental evolution. Change cannot be halted but it can, to some extent, be predicted and influenced. This is not a simple task because each object or collection presents a number of different factors in terms of raw materials, technology, condition, context, significance, current and future use. Given the possibility of change in all these areas, the conservator's task involves the need to identify and work with a large number of variables.

The analysis of condition, evaluation of current change, and attempted prediction of future rate and direction of change distinguishes the modern approach to conservation from the 'one-off' approach to intervention in the past.

Change induced by manufacture of objects

One set of variables that must be taken into account arises from the material nature of the object itself, so conservators need an understanding of raw materials and technology in order to assess the state of an object. For example, different methods of leather production produce materials with very different properties which will react very differently to conservation treatments. The behaviour of ceramics depends on their porosity (which depends on their manufacture), while the composition of a paint layer and the way in which it has been applied affects not just its durability but its reaction to methods of cleaning.

To help them interpret what they see, conservators and materials scientists must derive information on pre-industrial technologies from a range of sources. There are extensive reviews of ancient and pre-industrial technologies (Needham 1954, Singer *et al* 1954–8, Forbes 1955 onwards, Hodges 1964, Hodges 1970, Adam 1994, Nicholson & Shaw 2000). Other sources of information include early writings, probably of varying accuracy depending on whether the writer was a craftsman such as Cennini or Theophilus, or was simply reporting on the techniques, as did Vitruvius and Pliny (Humphrey *et al* 1998), observation of continuing (and evolving) craft traditions (eg Sturt 1923, Picton & Mack 1979), experimental work (Coles 1979) and, of course, investigation of the objects themselves (eg Tylecote 1962, Emery 1966, Picton 1984, Kingery & Vandiver 1986, Florian 1990).

Extraction and processing of raw materials

When assessing the state of an object prior to conservation the conservator should investigate the type of raw material and the way it has been used, because this has an impact on the significance of the object, and because the nature of the material affects deterioration and reaction to conservation treatment.

Extraction and modification of raw materials can involve a physical process such as mining, digging for clay, felling a tree or unwinding a silk-cocoon, or biological processes such as retting (controlled rotting) to extract plant fibres from the parent stem or leaf, or chemical processes such as tanning skins to produce leather. Shaping may involve cutting and carving (wood, bone), cutting and sewing (textiles, leather), weaving (basketry, textiles), modelling (clays, plasters) and so on. Decoration involves the application of surfaces such as paints, gilding, or the addition of features such as beads or inlay, or the modification of colour as in dyeing.

The skill of craft workers lay, and still lies, in their knowledge of materials and their behaviour, their control over modification processes, and their ability to produce a variety of products each, possibly, intended for a different purpose or with a different appearance. A very wide range of materials has been, and can be, produced, each with different characteristics and propensities to change and deteriorate. This means that, from the conservator's point of view, there is a huge range of different variables to recognize and deal with.

Raw and extracted materials were, and are, selected for their different characteristics such as their working properties, their potential for modification, their appearance and their durability. Thus lead may have been used because it is malleable; copper because it can be alloyed with other metals to produce a range of different products, such as bronze or speculum or brass; lime-wood (*Tilia* spp.) may have been selected for its ease of carving and lignum vitae (*Guaiacum officinale*) for its density and resistance to wear; gold because it was prized for its colour, shine, inert properties and rarity; and coloured marbles for their decorative appearance.

Some raw materials are relatively little changed in the process of making an artefact such as flexible woody stems used to make simple baskets. Other materials may be fundamentally changed, producing a new material with different properties from the parent material(s). Some of the most dramatic changes are induced by heat so, for example, soft and mobile clay is turned irreversibly into a hard and brittle ceramic by heating. Depending on the starting materials (clays, fillers, applied surfaces) the firing conditions (oxidizing or reducing) and the temperatures reached, the range of different end-products (all of them classed as ceramics) is enormous, and so is the range of treatments that they require. At the extremes, low-fired ceramics such as some British Bronze Age urns are weak, crumbly and highly porous, whereas high-fired stoneware or porcelain is hard, dense and non-porous.

The same raw materials may have been extracted and processed to different degrees for different purposes. Limestone required for building stone may simply be extracted from the quarry and shaped but not modified in other ways, whereas the same material may be used to make lime plaster by heating to break down its chemical structure, then mixed with water (slaked), and with sand or other filler, and finally applied to a wall as a plaster. After this processing the working properties and possible uses of the plaster are quite unlike those of the original stone. Some rocks are sources of both pigments and of metals. Iron-bearing stone can be used as building material, some forms can also be used to yield yellow or red pigments, but if iron-rich rock is further processed the material can be changed to produce a different substance, iron metal. Iron can be detected in the ore, the pigment and the metal but each looks and behaves differently. Similar variations in treatment can be seen in organic materials. Animal skin can be used whole as in raw-hide or it can be chemically modified to produce leather (in fact a variety of leathers from drum-skins to kid gloves). Hair can be used whole as in yarn, or it can be reformed as felt. Each product has different properties in terms of durability, strength, adaptability for different uses, and in terms of reaction to conservation treatment.

Construction of artefacts

Complexity

Objects can be of a single material and simple construction (such as a plain ceramic shape with no added surface or decoration, or a textile constructed of one type of undyed fibre used in a simple weave), or they may be made up of a number of different layers or components. Methods of construction add more variables to the structure of an object, including the way in which component parts are assembled, and the methods or materials used to hold them in position. The more complex the range and combination of materials and/or the construction, the more varied the changes resulting from deterioration and the greater the challenges for conservation.

Some objects are complex, consisting of a number of structural components and a range of added decoration, such as shoes or intricate metal objects. Understanding the complexity of construction and the number and nature of the components may thus be of fundamental importance to conservation and particularly to reconstruction of disassociated parts. The complexities may be apparent only to specialists who are familiar with the technology, and only then after some form of technical examination such as X-radiography has been undertaken. Identification of different components and their interactions is necessary in order to understand the processes and products of deterioration and to select a suitable approach to conservation. Applied surfaces or features may be firmly attached to the body of an object, such as paint with good adhesive properties, or a glaze fired onto the surface of a ceramic, but some elements may be only tenuously attached, eg beads sewn onto a textile. Some components require a third material to attach them to the substrate, such as binding, stitching or adhesives which may not be immediately visible (eg an adhesive holding veneer in place or metal dowel joining parts of a marble statue). These joining materials provide yet further levels of complexity, and potential threat to the survival and conservation of the object as they will have their own pattern of behaviour in terms of reaction to fluctuating environmental conditions, deterioration and response to conservation treatments.

Surface or substance

The character and significance of an object may be most obviously and immediately interpreted from the nature of its surface, planes and contours, colour and texture, but the nature of the surface is influenced by the structures and materials beneath. The ways in which the object has been shaped also affect the superficial appearance. Shaping which leaves tool-marks (eg in wood or metal) not only leaves evidence of the tool and the working techniques, but also a characteristic texture. A similar form may be created by pouring a liquid (eg plaster) into a mould but the resulting cast may have a very different surface quality from the original.

A good example of the complexity which may lie beneath a surface is that of easel paintings. The viewer examines and appreciates the image on the finished surface, but paintings are in fact highly complex artefacts, composed of many layers and components including the original canvas substrate, the stretcher, the frame, preparation layers, superimposed layers of paint, glaze and varnish, old restorations, relining canvas and adhesive, and so on. All of these components affect the character of the surface, and the role played by each of the layers must be considered by the conservator, as must the potential for damage to the surface through deterioration or the adverse reactions (and conservation treatment) of any of these layers.

Repair and reuse

Just like buildings, many objects have been modified during their life-times in order to keep them in use. Books have been rebound, polychrome sculptures and machinery, such as locomotives, have been repainted, objects have been repaired with solder, rivets, stitching and patches, costume has been dismantled and remade in other, more fashionable, shapes. Some material has been reused such as ground up ceramic used as 'grog' (filler) in a clay mix for a new ceramic, or mixed with lime plaster to induce setting in damp conditions. Materials from one context are used in another, such as European materials used in nineteenth- or early twentieth-century ethnographic objects. Conservators must detect these modifications when assessing the state of an object because they have an impact on significance, and may add yet more materials and variability to the original.

Imitations and copies

Some objects are made in one material but mimic another, as in prehistoric ceramics which are shaped and decorated to look like gourds or basketry (eg pottery from the Halaf period of Arpachiyah, Iraq, shaped exactly like basket work: see Hijara 1980).These imitations were not necessarily intended to mislead but may have been made for a number of different reasons, such as the continuation of tradition (baskets lined with clay are known from some sites such as Nahal Hemar (Bar-Yosef 1985) which might have led to baskets imitated in clay). They may have involved the substitution of less expensive for costly materials, or resulted from the discovery that using a different material meant more rapid production or a more durable result. Imitation can be found in many different periods and technologies, and ranges from the deliberate creation of 'gold' surfaces through manipulation of copper alloys in South America (Bray 1993) or use of yellow varnishes over silver (Portell 1992) to the imitation of ivory and tortoiseshell in early plastics (Katz 1984). Discovery of these imitations not only contributes to technological information but has an effect on significance. Some of these imitations are very similar in appearance to the material imitated but, of particular importance for conservation, they will behave in different ways to the originals in terms of deterioration and response to conservation treatments.

There are, of course, more recent deliberate fakes which have been coloured and textured to look like authentic ancient objects and which are almost always intended to mislead the acquirer and gain fame or financial advantage (or both) for the maker (eg see Jones 1990). In a reversal of the usual concept of faking, at least one unscrupulous dealer is known to have painted genuinely ancient objects to look like modern tourist art in order to export them from their country of origin with impunity (Watson-Smyth 2000a). On the other hand, copies made for didactic reasons may make no attempt to conceal the fact that they are made of a different material. Their intention is to provide information about form and surface detail, eg plaster casts of classical marble statuary, or the replica of the Anglo-Saxon (copper alloy) Sutton Hoo cauldron made of aluminium (Bruce-Mitford & Evans 1983). Some copies, particularly early plaster casts, have acquired a special significance because they record the condition of the original before subsequent deterioration, such as the casts in the British Museum which record the condition of some of the Parthenon marbles in the early nineteenth century (Plenderleith 1956). Casts such as these have achieved significance as early copies and are accepted as museum-quality objects in their own right.

Change induced by deterioration

Just as rock is slowly broken down and changed to a component of the soil, and plant litter is attacked by microorganisms and returned to the soil, so artefacts are subject to their own forms of deterioration. Whereas some raw materials, such as skin, are unlikely to survive unless processed, some are more resistant to deterioration in their natural state than in the modified state. For example, metal ores are more stable than the extracted and worked metal which reacts to surrounding factors such as water and oxygen to corrode and return to a state similar to the ore. Wood is more stable when the tree is still living than when the tree has been felled, and it is more stable as a felled tree if it is left in the round, since once sawn into planks the cell-structure is cut open and exposed to marauding and destructive organisms. Artefacts are made for use, and use will induce further change, from wear to major damage. Countless artefacts have crumbled and decomposed leaving little or no trace of their existence.

> *Ultimately ... they die, suddenly in accident, war or act of God, or gradually through the slow processes of decay. Their remaining and disjointed parts become again raw material, sometimes to be deliberately reused, like crushed ceramic utilised as grog in the making of new pots, and sometimes coming to degradation so complete that only elaborate scientific analysis could recognize in the residue the object that once was. (Pearce 1992, 16)*

The causes of deterioration may lie in the object itself, or the change may be promoted by use or by external agents such as water or insects. Before a conservator can choose a suitable conservation process it is essential to understand not only what the object is made of, but how and why each component has deteriorated. Understanding of deterioration mechanisms is fairly well developed for some modern materials which have industrial applications, such as synthetic organic polymers (Horie 1987), and for those with major implications for cultural heritage such as building stone (Amoroso & Fassina 1983, Price 1996). However, for some pre-industrial materials, particularly natural organic materials, comparatively little is known, and conservators and curators need to be aware of these gaps in knowledge. Much of the conservation literature focuses on deterioration in museums where the causative factors most frequently discussed are light, relative humidity, pollution and insects (Thomson 1986, de Guichen 1988, Koestler 1991, Pinniger 1991, Ginell 1994, Cassar 1995).

Rates of deterioration

Conservators must distinguish between change which has happened in the past and is now 'finished' (for example breakage and staining or build-up of foreign accretions), deterioration which is continuing, and the potential for renewed or different changes in the future.

In any one case several factors may be acting on the object, some more obvious than others, and this may make prediction of the rate of change difficult (Koestler 1994). A further complication is that deterioration does not progress as a constant but may proceed at different rates at different times in the life of the object. This is apparent when dealing with excavated objects which may have been deteriorating very slowly, or may perhaps almost have reached some form of equilibrium during burial. Excavation causes an abrupt change of environment and rapid deterioration can set in (Cronyn 1990). Glass may appear healthy but can, in fact, have undergone substantial chemical and structural changes (either during burial or in the atmosphere) with scarcely any visible degradation of the surface. It can suddenly become opaque if exposed to raised temperatures which cause the surface to crack or crizzle (Brill 1978, Newton & Davison 1989). Thus this difference in rate of deterioration may be inherent to the object, or rather to the material(s) it is made from, but it may also be due to fluctuations in external factors such as humidity or temperature, which is why not only excavating, but transporting objects from one climatic zone to another can be particularly hazardous.

Feller (1994) discusses the fact that deterioration may occur in stages and that there may be a comparatively long lead-in time (induction time) during which change may be imperceptible.

> *If conservators observe the condition of an artefact during an induction period and this period is a feature of its particular aging behavior, they may be lulled into thinking that changes are not occurring and that the artefact is in no imminent danger of change, only to find, a relatively short time thereafter, that decided changes have become apparent. (Feller 1994, 95)*

Feller was writing particularly about modern organic materials such as plastics and varnishes, which are of great importance to objects which have had modern adhesives or coatings applied during a conservation treatment, as well as to collections of early plastics. There is also much still to be discovered about how and why many older materials deteriorate. There is considerable interest in improving methods of detecting the slight changes, such as subtle alterations of colour and texture, which may herald later deterioration (Orna 1994).

The build-up of corrosion products to form a crust which may, or may not, be protective to the material beneath, is sometimes taken to indicate a slowing or halting of further deterioration. Whether or not such crusts in metals, in building stone or window glass are in fact protective is the subject of debate. The effect depends on a number of factors, including the porosity of the crust and the external environment. Some materials within crusts do remain unchanged for considerable periods of time; the prized patinas on some bronzes are an example (Schultz *et al* 1992). The nature and behaviour of the crusts, their aesthetic value and the evidence that they contain are important factors in determining whether to retain or remove them, and in estimating the short- and long-term effects on the stability of the object.

External causes of deterioration

Major natural disasters, such as earthquakes, storms and floods, can be responsible for dramatic damage, as can man-made disasters such as war-time bombardment (Fielden 1987,

Ashley-Smith 1999a) but the less perceptible and often cumulative effects of 'everyday' deterioration can be slowly devastating. This form of deterioration is normally described in terms of physical, chemical or biological mechanisms, or in terms of the effects of agents such as light (which in fact causes chemical deterioration). Many deterioration processes are further affected by temperature and by the presence or absence of oxygen. No single agent is likely to be responsible for the deterioration of a monument, collection or individual object (Brimblecombe 1994, Koestler 1994). Most types of deterioration require more than one agent; for example, deterioration of organic material by light almost always involves oxygen too, moulds normally need both moisture and warmth, some insects need the presence of fungi, and so on. Many objects will be affected by a number of different types of deterioration at the same time, thus stone buildings are likely to be affected by water, salts, organic growths such as lichens and a range of pollutants, textiles are commonly affected by both light and humidity, and metals by water and oxygen.

Physical deterioration results in cracks, chips, distortion or breakage, and is caused by impact or vibration, changes of temperature or humidity (which can cause some materials to expand and contract to a damaging extent), or faults of manufacture which may set up internal stresses in the object (eg Ginell 1994). Chemical deterioration usually involves a reaction with external agents such as water or pollutants and is most obvious in the form of corrosion on metals, but it can involve reactions, often caused by light, which depend also to some extent on internal factors as in the photochemical breakdown of paper, textiles and plastics (eg Thomson 1986, Reilly 1991). Biological deterioration involves living organisms such as bacteria, insects or plants, and almost always results in mechanical weakening and sometimes in complete destruction. It mainly affects organic materials such as wood, paper and textiles, but buildings constructed of inorganic materials can be affected by lichen or be disrupted by the root action of higher plants (Koestler 1991).

Water

Water is implicated in many kinds of deterioration. It is never pure but usually carries dissolved oxygen and salts and may be acidic to some degree. During burial the water regime in the soil has a profound effect on whether buried materials survive, or to what extent they are affected. In areas of relatively high rainfall and good drainage, the leaching out of soluble components of an object may result in rapid deterioration. Conversely, where water is absent buried materials may survive in spectacularly good condition. In areas of limited rainfall and rapid evaporation salts are not washed through the soil and tend to accumulate near the surface and in buried objects. In a situation where water is trapped and static, thus largely excluding oxygen (as in a well), organic material (wood, leather, some textiles) survive, although they are not unchanged. Oxygen is dissolved in free-flowing water such as in rivers, and in sea water and in these contexts organic objects are preserved only if they are buried in submerged silts where oxygen is excluded. Impressive examples of this preservation include ships such as the *Wasa* in Sweden, or the *Mary Rose* in the UK, in which the parts that were not buried in silts had completely decayed, even though they were under water, while the buried structures and the ships' contents were exceptionally well preserved (Pearson 1987, Sease 1987, Cronyn 1990, Watkinson & Neal 1998).

Above ground the water held as vapour in the atmosphere plays an important role in the deterioration of standing structures and of objects in museums. The key factor is not the absolute amount of water present but the drying effect of the atmosphere. This effect (known as the relative humidity, RH) is directly related to temperature: when the temperature rises the drying effect increases (that is, the relative humidity falls), even though the absolute amount of water remains the same. However, if the temperature falls the atmosphere loses its ability to hold moisture and the relative humidity rises. As the RH rises conditions may eventually become damp enough to encourage mould growth on objects, or even condensation of water droplets on cool surfaces (Thomson 1986).

Water is involved in many chemical reactions which result in corrosion of metals. Many of the salts deposited during burial, or penetrating into walls of monuments with rising damp, react with moisture in the atmosphere, resulting in volume change in the salts themselves which creates internal stresses in the object. Organic materials such as wood and textiles naturally contain water, and they gain and lose water in response to the relative humidity in the atmosphere, resulting in shrinkage and swelling, and stress. Fluctuating relative humidity can be particularly damaging since it causes repetition of stress, eg when salts repeatedly change state within objects, and where organics alternately shrink and swell. In conditions of high relative humidity moulds and fungi may grow on organic materials causing staining and mechanical damage (Thomson 1986, Staniforth 1992, Erhardt & Mecklenburg 1994, Scott G 1994, Erhardt *et al* 1997).

Light

The energy provided by both natural and artificial light is involved in many aspects of the deterioration of organic materials, and both visible light and high-energy ultra-violet light (at the blue end of the spectrum) are damaging. Paper, textiles and many dyes and paints are vulnerable, with the most obvious results being fading of some colours and weakening of structure (eg the fading and disintegration of the edges of curtains exposed next to a window). The natural colouring materials in woods are also affected, some fading and some darkening. Synthetic organic polymers (adhesives and plastics) are also vulnerable to the energy in light and may disintegrate completely through photochemical reactions. Here the first signs of deterioration are often seen as a slight colour change, usually yellowing. Light also encourages the growth of some organisms such as the algae and lichens which affect stonework (Thomson 1986, Michalski 1990b, Staniforth 1992).

Temperature

Heat, including the heat generated by the sun and artificial light, can have an important effect on deterioration processes, especially as most chemical reactions become more rapid as the temperature rises. The warmth required by visitors to museums and galleries actually increases the rate of change in objects such as paintings (Michalski 1990a). Heat directly affects relative humidity, thus high temperatures may result in low humidities and the drying out and shrinkage of organic materials such as wood (Erhardt *et al* 1997). Some organisms such as insects and fungi are more active at moderately warm temperatures (but can be killed by very high or very low temperatures). Very low temperatures encourage freezing (and thus expansion) of water in porous materials such as ceramic and stone, causing frost damage. Alternating extremes of temperature can cause harmful cycles of expansion and contraction in some materials.

Pollutants

Atmospheric pollutants are materials which are viewed as affecting the 'purity' of the atmosphere, such as sulphur dioxide which is largely responsible for acid rain. Pollutants in museums come from a variety of sources including the polluted outside atmosphere, building and display materials, other museum objects, and from substances such as domestic or industrial cleaning materials. Buried deposits may be affected by substances which leach into and pollute the soil, such as nitrogen-rich fertilizers, salt applied to roads in winter, and even acid deposition, but here little is known, as yet, about their extent and overall effects (Mattson *et al* 1996).

Outdoor pollutants which affect buildings and monuments, as well as the internal atmosphere of the museum, include not only sulphur dioxide but also nitrogen oxides (each of which can react with water to form acids), ozone and dust- or air-borne dirt (which can contain gritty particles, minute fragments of skin, carbon particles, oil droplets and bacterial or mould spores). Building and display materials including concrete, woods, wood-composites, adhesives and textiles can release alkaline particles, organic acids, formaldehyde and hydrogen sulphide. Air-borne particles can settle on any unprotected surfaces causing discoloration, mould-growth and abrasion, oily matter can adhere to any surface and dirty particles can penetrate into porous objects (Fig 12). Ozone (created in large quantities in highly polluted sunny conditions) causes fading and breakdown of organic materials. Acids can affect a wide range of materials in different ways such as fading of colours or structural weakening of textiles, leather, and stone-work or corrosion of metals (Thomson 1986, Perry 1990, Tennent *et al* 1993, Cassar 1995, Green & Thickett 1995, Blades 1998, Ganiaris & Sully 1998, Yoon & Brimblecombe 2000). Other materials such as pesticides could be considered to be pollutants in that they were often applied indiscriminately to objects in the past, particularly prepared animal specimens (taxidermy) and ethnographic objects, leaving hazardous toxic residues which may not be immediately apparent (Goldberg 1996, Johnson 1999).

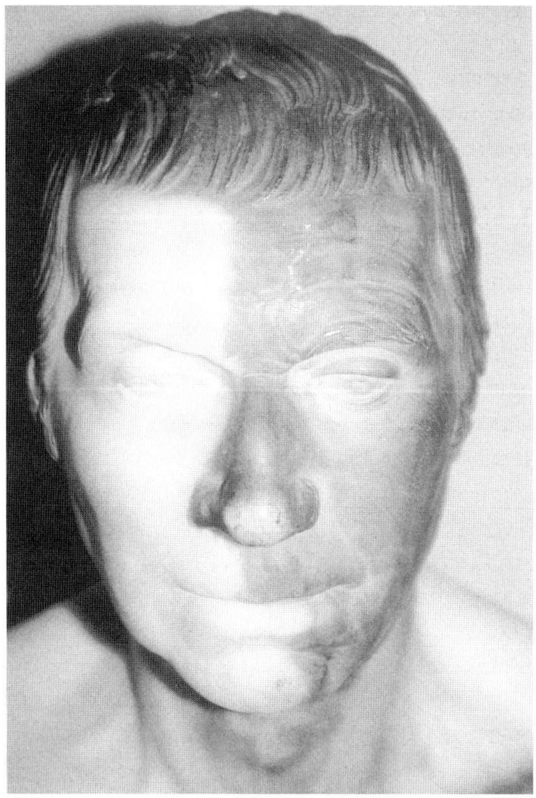

Figure 12 Pollution: marble bust after part-removal of oily dark deposits, which are typical of objects exposed to a city atmosphere (Courtesy of Museum of London).

Almost certainly, hitherto unnoticed contributory pollutant factors will be recognized in the future (Brimblecombe 1994), and there will be new factors to contend with, just as formaldehyde became a 'new' problem when reformed wood-composites such as ply and MDF (medium density fibreboard) became more widely used in construction in museums (Ganiaris & Sully 1998, Thickett 1998).

Living organisms

Organisms such as bacteria, fungi and insects can cause considerable damage through feeding on and weakening the object itself (eg the clothes moth, *Tineola bisselliella*); others such as marine borers (eg the teredo or shipworm, *Teredo navalis*) or higher plants may penetrate the structure and weaken it, or disrupt it through root action. In museums the effects range from the devastation caused by insect infestations, especially in tropical or semi-tropical conditions (eg termites, *Isoptera*) to the limited disfigurement caused by moulds, or by insects such as silverfish (*Lepisma saccharina*) which are selective feeders. Although organic materials are chiefly affected, stone can be damaged by algae and lichens and in buildings such as churches where it is difficult to exclude birds and bats their excreta can stain paintings and textiles or have corrosive effects on metals and polished surfaces (Koestler 1991, Pinniger 1991, Paine 1993, Scott G 1994, Urzi & Krumbein 1994).

Human causes

Although deterioration is frequently described in terms of physical, chemical or biological mechanisms, this masks the fact that much of the damage to, and deterioration of, the cultural heritage is caused directly or indirectly by human agency, by factors ranging from vandalism or war, to over-visiting or neglect (Keck 1976, Hodkinson 1990, de Guichen 1991). This is reflected in the various international codes and charters which refer to the problems of looting and other effects of armed conflict (eg UNESCO 1970) and in the current studies of the effects of over-visiting (eg Lloyd & Mullany 1994). Perhaps just as damaging are the direct and indirect effects of 'legitimate' activities such as agriculture, building works and industrial processes, as well as the more insidious effects of poor practice amongst the people who interact professionally with the cultural heritage, including conservators. When reviewing condition of objects in museum storage, for example, it is not unusual 'to identify physical handling as the main cause of damage to the collection, a problem made worse by overcrowding' (Bradley 1997b, 6).

Structures and objects that remain buried are subject to the effects of human activity including pollution; physical destruction from quarrying, deep-ploughing, forestry and the digging of deep foundations for new buildings; deterioration resulting from changes in the water table; or crushing resulting from increased loading on the surface such as might be caused by heavy traffic on a road (Coles 1987). Human intervention in the form of excavation then exposes objects to new conditions which can instigate changes such as flaking or rapid corrosion (de Guichen 1984, Cronyn 1990).

Human agency is emphasized by de Guichen (Fig 13) in his analysis of deterioration, in which he focuses just as much on human causes as on natural environmental factors. His analysis also classifies types of deterioration (caused by both human and natural factors) according to their rate of action, distinguishing between immediate and drastic causes (war,

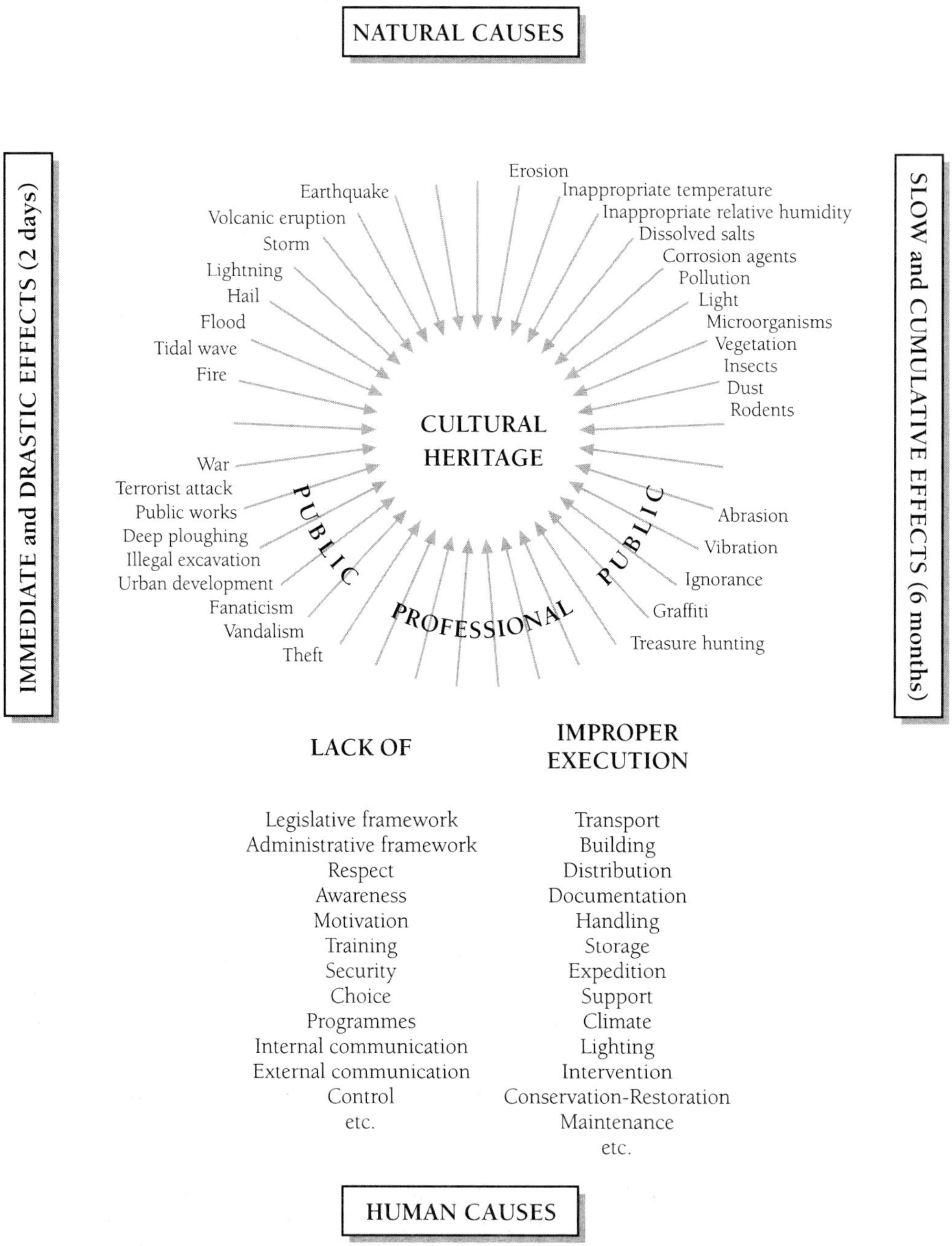

Figure 13 Causes of deterioration (Courtesy of Gaël de Guichen).

vandalism, flood and fire) and slow and cumulative effects (poor storage, faulty restoration, dust and microorganisms). This analysis emphasizes the very large range of variables which affect the way in which cultural material deteriorates and the many factors with which archaeologists, architects, curators and conservators have to contend.

Deterioration during burial

Burial exposes objects to the physical effects of the overburden of collapsed walls or soil build-up which can cause crushing, flattening or distortion. The nature of the soil (acid or alkaline, free-draining or impermeable) together with the presence or absence of water, oxygen and microorganisms may cause either relative preservation or total decay. Most organic objects, such as wood, leather or textiles, simply disappear in well-drained damp soils. Indeed, in most climatic zones burial presents 'ideal' conditions for decomposition of organic materials. Frequently the archaeologist and conservator can hope to find only partial evidence of organic materials, such as mineralized traces preserved in contact with metals (Janaway 1989, Gillard *et al* 1994).

Organic materials survive well only in exceptional circumstances such as very dry conditions (hence the excellent preservation on many Egyptian sites), in completely silted and oxygen-free waterlogged conditions or in permanently frozen environments (the Copper Age man preserved in the European Alps, Spindler *et al* 1996). Ceramics usually survive comparatively well but may be affected by salts carried in the soil water. Both metals and glass commonly survive but undergo chemical changes which result in conversion of part or all of the object to a different corroded form (Fig 14), frequently resulting in change of colour and/or dimensions.

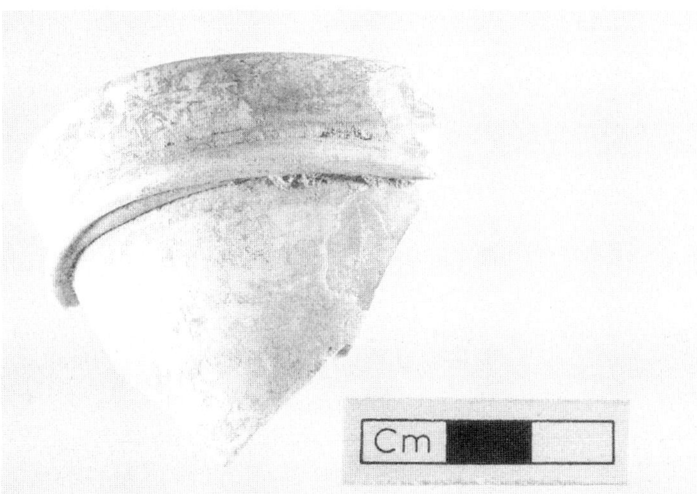

Figure 14 Fragment of Roman glass vessel showing the opaque crust which can form during burial (Courtesy of UCL Institute of Archaeology).

Problems inherent in the object

Further factors that affect deterioration may be inherent characteristics of the object and the materials it is made of; some materials are essentially more unstable or more vulnerable than others. Conservators sometimes refer to this characteristic as 'inherent vice' (Keck 1976, 331).

For example, with the exception of gold (and some others, such as platinum), most common metals corrode fairly readily in the presence of water and oxygen; or rubbers used in ancient artefacts, as well as in modern objects such as space-suits, deteriorate readily on exposure to light and oxygen and lose their elasticity (Baker 1995). Plastics are generally vulnerable, and some such as cellulose nitrate (eg photographic film and early plastic objects) hasten their own deterioration: as cellulose nitrate breaks down it produces acid which promotes still further decay (Blank 1990, Orna 1994, Pullen 1999).

Materials which are porous (some types of ceramics and stone, and organics such as wood) are vulnerable because water, salts, mould and fungal spores etc can enter the pores and cause various forms of damage. Thus, even though the substance of a ceramic is relatively inert, if the structure is porous it can be badly affected by the growth of salt crystals. Organic materials are responsive to changes of humidity, so textiles and wood continue to expand and contract unless the external conditions are steady and unchanging (Hoadley 1978, Erhardt *et al* 1997), and they are vulnerable to attack by insects and fungi (unless treated in the past with a toxic pesticide). Many pigments fade or change on exposure to light (Michalski 1990b). Some materials (eg waxes) soften in heat, some (eg shale or ivory) have a characteristic layered structure and a tendency to separate into these layers when weakened.

Physical damage can be caused by juxtaposition of materials which expand and contract at different rates and to different extents, as for example with a relatively immobile paint surface (which may become more brittle with age and change) applied to a wood substrate which regularly changes dimensions. Similar problems can occur when related materials are fixed in a structure, such as wooden components of a piece of furniture. The strongly directional properties of wood (it is anisotropic, expanding and contracting to a far greater extent across the grain than along the grain) can result in damaging stresses where components such as supporting rails and legs of chairs or tables are joined, or where decorative veneers are applied to a carcass of different wood. Other problems can be caused by juxtapositioning, for example where a metal which is relatively reactive in chemical terms (such as iron) will corrode even more readily when in contact with a less reactive metal (such as copper).

Interaction between different materials used in the manufacture of an object may cause deterioration. Some additives cause degradation of otherwise more stable materials; for example, in the late nineteenth and early twentieth century tin salts were applied to silk to produce a heavy luxurious effect but these weighted silks frequently disintegrate into brittle shreds (van Oosten 1994). Parts of the construction of an object may deteriorate and threaten the whole: corrosion of metal fittings is frequently responsible for staining and damage to other materials. For example, Anglo-Saxon antler or bone combs are made of several components which are normally held together with metal rivets. The corrosion of the rivets causes darkening and cracking of the bone or antler which would probably survive well without the presence of metals (Fig 15). Enamel decoration on a metal substrate may be disrupted, discoloured or completely masked by corrosion of the metal (Hunter & Foley 1984). Similarly the weakening of canvas in an easel painting puts the painted surface at risk, and the failure of adhesives may threaten the construction of objects such as furniture. Some objects have had materials added as part of a conservation treatment and these too may cause problems, for example by producing residues (eg added pesticides) or by shrinking (modern adhesives or fillers used for repair).

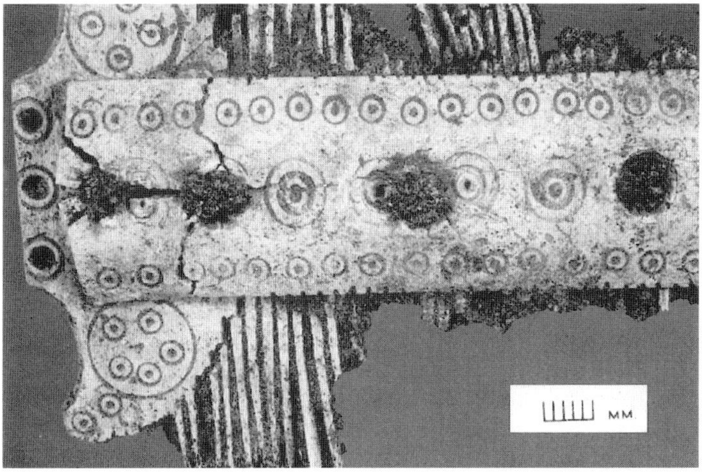

Figure 15 Inherent vice: medieval antler comb showing that the corrosion of the iron rivets has caused splitting of the antler (Courtesy of UCL Institute of Archaeology).

Faults in manufacture can also cause problems as when a ceramic glaze does not 'fit' the body of the ceramic, and the two do not expand and contract at the same rate, or when glass is insufficiently annealed and thus contains in-built stresses which may cause it to shatter. Impurities or lack of homogeneity in the raw material, or imperfections in technique, can also cause problems, as when a lime plaster has been poorly prepared and has failed to set completely.

Surface accretions, such as corrosion products of metals, have formed as a result of reaction between the object and the external environment and they are effectively part of the object. Other accretions may be foreign, in the sense that they are external substances such as salts or soil which have adhered to the object. If clarification of the appearance of the object is an issue, foreign materials are normally recorded and removed (and they would normally be removed if they are causing continuing damage) but decisions on inherent accretions are more difficult. These corrosion layers will have been formed at the expense of the material of the object itself and in some cases they preserve the original shape and dimensions of the object (or sometimes of materials that were in contact with the object). The crust may be retained, or only partially removed, where the object's original surface is within the thickness of the corrosion layers, as in many excavated bronzes. A particular problem is posed by ancient deep red glass where the red colour was achieved by the addition of copper salts. During burial not only may a corrosion crust form which obscures the red but the crust can be bright green as a result of reactions with the copper. This situation provides both curator and conservator with a dilemma since the shape and dimensions may be preserved by the crust, but the original colour is both masked and distorted (Smith 1999).

Effects of use and care of objects

Both previous and current use have an effect on change and deterioration. An object's previous use(s), or periods of neglect, will affect the current state and its future deterioration.

Any object or structure may continue to fulfil its intended role (eg a chair, or a dwelling house) for all its functional life, and this use may expose it to considerable wear and stress and the vagaries of modification and maintenance. Some things, however, may have a limited use and spend much of their time in storage, either because of their particular purpose (eg an 'heirloom' such as a family christening robe) or because they are now considered curious rather than functional (eg a veteran car). These may be more carefully protected than a garment or vehicle in regular use.

Museums usually hope to acquire objects which are in reasonably good condition but the older an object the more likely it is to have been affected by some form of change (and excavated objects are often, inevitably, in an unstable condition). Objects which have been given some special value are likely to have been better cared for, and so in better condition, than mundane objects such as working clothing, or tools such as saws and chisels (which may have been repeatedly sharpened until the metal is partially worn away). However, because fewer of these mundane objects survive, even examples in poor condition are of interest to museums.

On entering a museum, objects will be exposed to a new and different conditions, and used in a number of different ways which will expose them to changes of environment, all of which may be stressful for an object already weakened through deterioration. Handling for study, photography, conservation and display all put strains on objects and the movement between differing micro-environments is potentially damaging (Cassar 1995). Making objects accessible through display brings with it the need to position and light them in such a way that they can be appreciated. Working objects such as clocks may be regularly wound and kept going so that their function can be demonstrated; they can be particularly subject to change since the need to keep them in working order may involve replacement of moving parts. Viewing alone may not provide enough information, or may be inappropriate for some people. Thus some objects will be handled and manipulated by researchers, by visually impaired people, or by children: in fact some objects are used purely as handling or reference material.

Significant objects are often much in demand, so are made more widely accessible by being lent to other institutions for temporary exhibitions, exposing them to the stress of packaging, travel and change of environment. These same objects may be used to generate images or replicas which enable visitors to take a memento home with them, another form of use. For this they may be photographed, which involves handling and exposure to photographic lights (which may be damaging), or a cast may be taken from them, which means applying release agents and moulding compounds to the surface.

The context in which objects are housed affects the environment and thus whether and how they deteriorate. Large numbers of museum objects are kept in store and used only occasionally for research or exhibition. In the past when the importance of good storage conditions was less emphasized, many objects such as costume or basketry, for example, were folded or crushed into inadequate spaces, or exposed to air-borne dust. Placing different material types (ceramic, metals, bone, wood) in the same display case or storage container can lead to difficulties in that each material may need different conditions for stability, or may affect another adversely (metals should be kept in dry conditions to prevent corrosion, whereas wood may shrink and crack in the same dry conditions); substances in some materials, eg textiles or 'fresh' wood may cause metal corrosion.

Outdoor conditions are less controllable and tend to be more damaging than indoor ones so exterior stone and metal statues may deteriorate rapidly (and may eventually be brought into a museum to 'save' them, even if it means removing them from the original context). Equally, displaying an object in the open in a museum rather than in a closed glass case exposes it to dust, to the problems of maintenance cleaning and possibly to handling. Any change of environment can pose problems. Objects such as seal-gut parkas were made to be used in cold and wet arctic conditions and tend to become dry and brittle in a centrally heated and frequently rather dry museum environment (Morrison 1986); conversely semi-tanned skin products which are stable in the hot and dry conditions of the desert may suffer in temperate regions. Abrupt changes of environment can be very damaging, as might happen if a wooden statue or piece of furniture was moved from a poorly heated and rather damp medieval house or church to a centrally heated museum. The wood would have acclimatized to a fairly high relative humidity in its original context but when taken abruptly into much drier conditions would probably shrink and warp. To be successful this kind of transition has to be made by changing the environment slowly and gradually.

Restoration and conservation

Early restoration or conservation treatments may have caused deterioration, even though most of them were carried out with the best of intentions. Paintings on canvas are vulnerable to deterioration of the canvas support and the painted surface, and most paintings have had some form of restoration if they are over a century old. Some of this restoration has caused damage, such as removal of original paint or glazes through over-cleaning, flattening of surface texture or even imposition of an alien texture (Hodkinson 1990). Broken ceramics from, for example, Greek vases, are often found to have had the edges of the sherds ground down to make the fragments fit together more easily during reconstruction. Stone and ceramic objects have been repaired using metal dowels and clamps which not only involve drilling into the original object but may create further problems if the metal corrodes. Fragile textiles may have had new textile supports sewn to them and the stitching may be responsible for damaging stresses on the original. Metal objects may have had the corrosion stripped from them using acids or similarly rigorous treatments, or they may have been regularly polished (particularly silver) to retain a shiny appearance: in each case there may be loss of surface detail. Repainting of polychrome sculpture may mask the original detail and colour scheme.

Even recent conservation has been responsible for damage through the application of synthetic organic polymers in the form of adhesives or consolidants. When first used the long-term properties of these polymers were not well understood. Many of these adhesives or consolidants have changed or deteriorated themselves and caused damage through shrinkage, yellowing or embrittlement (eg Sease 1981), and deterioration and failure of an adhesive may put an object at risk. The ability to examine structures in ever greater detail (for example with scanning electron microscopes) has now shown that apparently beneficial treatments such as washing of textiles may cause swelling and damage to the fibres at a microscopic level, damage which is not evident to the naked eye (Cooke *et al* 1996).

Intended deterioration

Some objects are not intended by their makers to last. Ephemeral items such as bus tickets are not designed to be durable and are normally simply discarded, but the very processes of deterioration are considered to be an important part of the meaning of some artefacts, including some ethnographic objects (McLeod 1995) and examples of contemporary conceptual art (eg Magor 1990). Albano set out the views of a contemporary artist, Kiefer, a follower of Beuys:

> *In his view, change is all part of the process, and those who are concerned about his work deteriorating simply do not understand the meaning of his work. Beuys helped instill in*

Kiefer the perspective that the element of change was a natural and inevitable process and perfectly valid as an aspect of a work of art. (Albano 1996, 182)

Effects of deterioration

Surface or substance

The character and meaning of many objects is carried on their surfaces, and much deterioration affects the surface rather than the substance or body of the object (Brimblecombe 1994). This is particularly true of wear and tear, natural erosion by water or wind, mould growth and changes which depend on external factors such as pollutant gases and air-borne particulates. Painted surfaces fade or change on exposure to light or the paint flakes away from the substrate (Fig 16). Both metal and glass corrosion are normally fed by moisture and other external factors, starting on the surface and proceeding inwards. Water-soluble salts in porous materials such as stone or ceramics tend to crystallize where evaporation takes place, at or just beneath the surface, which causes the surface to flake or spall. Wood exposed to the weather as in buildings is affected by water and photochemical changes resulting in surface erosion.

Figure 16 Roman Mummy portrait showing loss of paint from the surface, and cracking of the now very fragile wood on which it is painted (Copyright: Petrie Museum of Egyptian Archaeology, University College London, UC 38103).

On the whole, materials of plant and animal origin are less likely to be affected solely on the surface. Solid materials, such as wood or ivory, split, sometimes right to the heart of the object, and distort when exposed to major changes of relative humidity. Sheet materials such as textiles, paper and some forms of leather (and plastics) can crumble and disintegrate completely as a result of photochemical deterioration. Solid wood attacked by insects can survive as a weak but intact skin over a core which has been reduced to powder (this effect is typical of termite attack in the tropics). Similar effects can sometimes occur when wood is attacked by wood-rotting fungi.

Major impact damage or physical overload can, of course, fracture or destroy many types of material, and badly broken objects tend to be uninformative. Inorganic materials can also be affected structurally rather than superficially, because many processes which start at the surface eventually destroy the whole object, as in some forms of metal corrosion, or when stone is attacked by acid rain.

Surface deterioration often results in a change of appearance and sometimes this change reduces the legibility and obscures the significance of the object. Objects can be stained by other substances. Bronze becomes green, silver becomes black. Yellowed and darkened varnishes obscure both detail and colour in paintings, corrosion obscures inlay, enamel and other decorative features on metal objects (Fig 17). Flaking and spalling of stone surfaces results in loss of sculptural form.

Loss or distortion of information

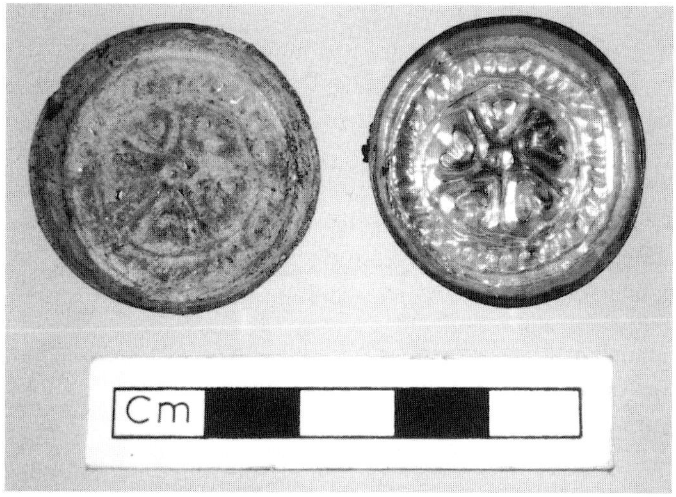

Figure 17 Pair of Saxon saucer brooches, one before (left) and one after (right) removal of obscuring corrosion products. When first excavated the gilded surfaces were not visible (Courtesy of UCL Institute of Archaeology).

An important distinction to make between buried and other artefacts is that in many cases excavated objects have survived only partially. For example, metal tools survive where their handles do not because the burial conditions have caused the wood or bone handle to decay. This propensity of organic materials, such as wood, skin, textiles, to decay causes a bias in many collections of excavated objects in favour of inorganic materials such as ceramic, stone and metal. Some idea of the range of material which has been lost can be gained by looking at ethnographic collections which show that materials of plant and animal origin can be (and probably were) used to make a wide range of objects, even before ceramics and metals were manufactured.

There can, of course, be spectacular preservation of organic materials from desiccated, waterlogged or frozen burial conditions, but these are exceptional. Even in conditions which are apparently conducive to preservation of organics, there may be some distortion of information because the soil conditions favour preservation of some organics but not others. This can happen where the soil conditions are acid enough for fine plant fibres such as linen to deteriorate without trace, even though wool survives. (In these circumstances survival of wool textiles is not a reliable indicator of the range of possible textile types that might have been in use.) Part of the fascination of material from Egyptian excavations is that much, or all, of what is missing on temperate sites is preserved, including clothing, furniture and all the paraphernalia of daily life.

Another form of distortion of evidence can result from the addition of materials during conservation treatments which change the appearance or composition of the object. Treatment of bronzes with sodium sesquicarbonate (in fairly common use until the last quarter of the twentieth century) has left an unnaturally vivid blue-green patina. Electrochemical reduction of metal corrosion involving the use of zinc granules or powder and an acid (also in fairly common use as a treatment until the last quarter of the twentieth century; see Plenderleith

1956) may have left zinc on the surface of the object and so make a genuinely ancient object appear to be modern and possibly a fake (Craddock & Bowman 1991).

Limitation of use

There are strong links between condition and use. If an object is in poor condition this obviously limits its use. Sometimes condition makes a particular use impossible, as when a request to lend an object is refused on the grounds that the condition will not allow it to travel. Through improving condition, conservation has a role in making objects usable, or keeping them in use. Occasionally the condition is so poor, and the possibilities of improvement through conservation so limited, that an object will be considered to be unusable and will be subject to disposal.

Attitudes to change and deterioration

Without change and deterioration the world would be cluttered with objects and structures of all kinds and ages, creativity might be stifled or at least constrained and there would be no need for conservation. 'If everything from the past had been permitted to survive, what other problems might humanity have today?' (Keck 1976, 337). 'God created the world and all things in it in six days; on the seventh He saw how they proliferated; on the eighth He created moth and rust' (Lowenthal 1994, 41). Deterioration has resulted in the loss of large amounts of cultural heritage. Nevertheless, as death has influenced some of the most beautiful of art forms, so deterioration can be seen as a significant and shaping force in our attitudes to the heritage and how we use it.

Change and history; 'pleasing decay'

Deterioration provides the convincing signs of age which are valued as evidence of the life of the object or as adding to its aesthetic appeal. That a thing is old gives it a particular significance and the effects of use, damage and deterioration are valued as evidence of the life-history of the object (if the object were new these same effects would probably be considered out of place and unacceptable). Some forms of damage caused by human actions are historically significant, so are viewed more as evidence than damage, such as early annotations to manuscripts, or even the owner's doodles on an early school book: even though they may have degraded the object their effects would not normally be modified or removed (see Fig 8). Examples of obviously significant damage include the rents and blood-stains in the Archduke Franz Ferdinand's uniform caused by his assassination in 1914 (which launched the First World War), or damage caused during events such as the French Revolution of the late eighteenth century (Page 1999).

Wear marks provide evidence of continued use, so it is difficult to believe that some objects which show little sign of age (apparently pristine gold objects, or objects which have been thoroughly cleaned or restored) *are* really old. The conviction that 'new' materials are long-lived and immune to decay (eg space suits) may have slowed the realization that they are also vulnerable and need appropriate conservation measures (Baker & McManus 1992). The effects of age imply longevity – that the object has lived through and been involved in the 'past' – and they provide the 'cachet of antiquity' (Lowenthal 1994, 43). Some of the adverse reaction to cleaned paintings lies in the fact that the paintings have lost their familiar appearance of age and the newly exposed brilliant colours cause considerable shock. Signs of age are considered important enough for age to be falsified by 'distressing' and other techniques applied to modern furniture and clothing (Lowenthal 1985).

Signs of age may add to aesthetic appeal. 'A patina of age enhances the perceived value of many artefacts, notably buildings and works of art. Age and decay gained cult status in Western civilisation with the Romantic taste for ruins and fragments' (Lowenthal 1994, 39). Yellowed varnish mellows the tone of a painting, corrosion can form a desirable patina and crumbling stone contributes to the atmosphere provided by partially ruined buildings.

> *Some of these alterations are sublime. To that beauty imposed by the human brain, by an epoch, or by a particular form of society, they add an involuntary beauty, associated with the hazards of history, which is the result of natural causes and of time. (Yourcenar 1996, 213)*

The products of deterioration can also be informative. Excavated metal objects may retain traces of textile or leather 'fossilized' in their corrosion layers and corrosion may retain original surfaces and dimensions. These products may also indicate whether an object is genuinely old or not, since it is difficult to replicate corrosion and other aged surfaces. A yellowing varnish may not only provide a mellow appearance but may help to retain the colour balance of a painting (nearer to the artist's intent) where greens have faded to blue.

This value put on the effects of deterioration presents the conservator with a challenge because his/her role is to manage ongoing change, while at the same time ensuring that the desirable effects of past change are unaffected. This challenge has been met effectively by the National Trust at Calke Abbey, UK, where the signs of the decline and eccentricity of the house's final years in family ownership were carefully preserved (Drury 1988), by English Heritage at Wigmore Castle where the appearance of a romantic ruin was retained (Coppack 1999) (Fig 18), and at Brodsworth Hall where the decision was made 'to tell the history of the house, including the undeniable effects of time and neglect' (Allfrey 1999, 117).

Figure 18 Wigmore Castle, Herefordshire, UK. Conserved as a 'romantic ruin' (Courtesy of N D Balaam).

Change and decrepitude

However, not all deterioration is desirable. Decay can go beyond the point of being pleasing and can become decrepitude. This is reflected in the fact that several forms of deterioration are commonly described as if they were human illnesses (sick glass, bronze disease, warty corrosion) or some other form of unpleasantness (red rot of leather, weeping iron, weeping glass, tin pest). Extensive damage may mean that the object is unrecognizable and its significance is not just obscured but is reduced or negated by loss of meaningful form or surface features: it is no longer understandable or usable. Thus buildings pass from being picturesquely ruinous to being meaningless rubble. Conservators and conservation architects must aim to manage change so that continuing deterioration does not reach the point at which it threatens the integrity and significance of the object.

Managing change

Change is an inevitable consequence of *using* the objects and structures which make up the cultural heritage. Therefore the conservator's task is to work with change, to accept it, to understand it (as far as this is possible), and to manage it.

> *It goes without saying that ... this great heritage cannot be embalmed, frozen, held in a state of museological stasis. Change is ineluctable in both artefact and context. But the rate and nature of this change can and must be controlled.* (Fitch 1982, 350)

In the past the craft restorer's intervention was often a one-off and there was little consideration of reasons for change, or of preventing future change. Now the conservator aims to identify the large number of variables which affect the state and use of the object and to decide, with the stakeholders and other interested people, what measures to take to moderate existing and future change. The variables lie in the composition of the object just as much as in the factors which may have caused it to deteriorate in the past, and may cause further deterioration in the future. They also lie in the varying interpretations of the object and the shifts in its value.

The Delta Plan provides an excellent example of conservation as management of change (Buisje 1995, Talley 1999). It was set up in The Netherlands in response to what were perceived to be unacceptable rates and extents of deterioration within state museum collections. It involved not only changing the priorities of politicians but adopting a changed approach to conservation, classifying and selecting material for attention, concentrating on preventive measures, doing no restoration and involving volunteers to work with conservators. On a smaller scale, individual museums and other heritage organizations are adopting a management approach to conservation, including environmental management, collections management, visitor management, integrated pest management, indicating that conservators and curators are accepting and working with change in a positive way (Cassar 1995, Keene 1996, Pinniger & Winsor 1998).

Change and conservation

Understanding of changes the object has undergone during manufacture, use and deterioration enables the conservator to attempt to predict not just future changes but the consequences of his/her actions. This means it is possible to begin to develop a conservation strategy rather than simply react to immediate and obvious problems. However, predicting change is not necessarily straightforward since materials and systems can behave in unpredictable ways. Chaos is just as likely to affect changes in objects as changes in the weather (Seeley 1999). At least conservators are now in a better position to consider a range of possibilities before conservation, and to monitor many of the effects of change following conservation procedures.

6 Issues in practice: assessment and decision

Introduction

This and the next chapter explore the issues related to the practice of conservation, taking this to mean day-to-day interaction with objects and a direct concern for their care and conservation. As each object passes through a range of different changes during its life, by the time it reaches a conservator it presents many different material and conceptual problems. After assessing the object and the factors affecting it, a conservator must identify the problem(s), establish the objectives for conservation, and evaluate options, in order to arrive at a decision on how to proceed. This chapter introduces the management framework within which conservation is undertaken and explores the processes of assessment and decision-making.

The management framework

Conservation plans and policies

Conservation should take place within the framework of policies or plans agreed for the management of monuments or collections. These are more or less detailed and exist at different levels depending on the size and complexity of the site, monument or museum.

For sites and monuments it is normal to develop a *Conservation Plan* which is, in effect, a statement of policy (see Semple-Kerr 1996, HLF 1998, Clark 1999). For museums and collections the overall framework for care of collections is the *Collections Management Policy* which covers the general issues involved with safeguarding and managing collections (Edson & Dean 1994, Fahy 1995, Lord & Lord 1997). Within collections management there may be other policies for environmental management or pest management (Cassar 1995, Pinniger & Winsor 1998).The principles set out in the *Burra Charter* (Australia ICOMOS 1999) are widely used in the preparation of conservation plans for monuments, but Townley and Parris (1994) have shown that similar principles can apply very effectively to collections, thus making a link between moveable and immoveable heritage.

The starting point for developing policies is to reach an understanding and definition of the significance of the monument, collection or object, and to establish in what ways it may be vulnerable. The plan or policy is intended to facilitate the management of the monument or collection so that the significance is not diminished or lost. Thus it sets out general policies for how best to achieve effective overall care. The information needed to draw this up is derived from written records, the views of interest groups or stakeholders, examination of the condition of the monument/collection and the environment in which it is situated, consideration of the ways in which it is, or can be, used, and consideration of the implications of its significance. The plan or policy will also take account of external constraints such as government policy.

For any one monument or collection, the policy may set out the 'vision' or the 'mission', as well as the way in which the monument or collection will be used and guidelines on matters such as appropriate conservation processes (eg whether sampling is permitted for scientific investigation, the extent of allowable restoration, or the treatment of 'working' objects), the type of documentation required, preventive care or maintenance and how to deal with natural or man-made disasters, or (relating to collections) acquisition, loans and disposals.

Policies are distinct from work plans, which detail how to get things done. These might include project plans, collections management plans, programmes of work etc. These plans set out the procedures and timing of the conservation work itself; they might give details of how the condition of collections will be improved in stages over the next five years (Keene 1996), or details of a conservation project which is going out to tender (Conservation Forum 1998). They may detail not only processes and timing, but also cover aspects such as the equipment needed, the personnel involved and so on.

The aim of both conservation policies and plans is the management of change. The policy provides guidance on the overall preservation standards and philosophy, whereas the project plan or programme of work details what is intended to happen to bring a specific project to completion.

Collections management

Collections management policies provide the overall framework within which all aspects of the conservation and use of museum objects take place. These policies have made a major impact on the condition and care of collections. This form of management is concerned with every activity which affects objects, so depends on establishing the composition, significance and state of collections, and the environment(s) in which they are held. Because it is so wide-ranging and affects so many people who interact with collections, this form of management often embraces the education of staff and other stakeholders in order to ensure that they can play their part in establishing and maintaining effective conditions for preservation. It includes the development of policies and practice for controlling environments and for dealing with issues such as security, acquisitions, loans, transport of objects, de-accessions, and disaster such as fire or flood. For some, the term 'preventive conservation' has widened in meaning to embrace many of the activities listed here as being typical of collections management (Williams 1997, de Guichen 1999). In that it is the business of museums to preserve collections, perhaps 'preservation management' better explains what is embraced by the term collections management.

Figure 19 Demonstration display case showing hinged door, separate vented light box and use of aluminium barrier films to protect objects from pollutant effects of fibre-board carcass: the standard adopted by the Museum of London for permanent displays (Courtesy of Museum of London).

Management of the environment is an essential tool in the management of collections in that it contributes to maintaining good conditions for the preventive care of objects (Fig 19). The monitoring and control of factors such as light, relative humidity, temperature, pollution and insect pests, all of which affect the conditions in which collections are housed and used, play a crucial role in managing change and thus in long-term conservation (Thomson 1986, Knell 1994, Cassar 1995, Keene 1996, *Museum Practice* 1996 onwards, Pinniger & Winsor 1998).

The advent of collections management and the consequent better knowledge of the state of collections has been responsible for encouraging re-evaluation of the risks objects are exposed to and the development of new guidelines and more flexible approaches, for example to the control of light or relative humidity exposure (Michalski 1990b, Erhardt & Mecklenburg 1994, Ashley-Smith 1995, Keene 1996). Better monitoring of collections has also increased the amount of planned or proactive conservation treatment of individual objects, and ensured that potentially serious threats such as a pest infestation are caught at an early stage. Thus collections management policies play an important role in safeguarding condition and significance of objects and collections.

A current issue which affects all museums (and monuments) is the requirement to provide greater access to the public, and here management policies are essential to enable museums to make sensible decisions about the use of collections. There is increased pressure to put on major exhibitions, to open storage areas, to demonstrate working objects, to encourage handling of objects and so on. This is a challenge to conservators who have to ensure that objects are usable, while at the same time attempting to conserve them for the future. Getting the balance right can be difficult.

> *Faced with the need to use and interpret art for the present generation, we can lose sight of our moral obligation to pass it on to the future, with its physical and metaphysical significance undiminished at least, and enhanced at best. (Hodkinson 1990, 60)*

Management policies which tackle questions such as whether machinery or musical instruments should be demonstrated, and if so how often, and how they should be maintained, aid conservators and curators in reaching appropriate decisions on how to balance use today with survival tomorrow (Keene 1994, Mann 1994a, Mann 1994b, Barclay 1997). In other words, collections management policies facilitate the use of objects within the context of long-term preservation.

Managing the environment is specifically concerned with the building itself (or micro-climates within it), its structural integrity, ways of controlling ingress of water, light, pollutants, insects and so on. It also concerns factors such as opening times and opening periods. For example, one way to control the length of time objects are exposed to light is to close buildings to visitors for set periods of the year. In the UK, many National Trust properties are closed during the winter months which also removes the need to heat the houses for the comfort of visitors (which could involve temperatures which are not necessarily suitable for the collections).

Environmental management deals with the use of space within the building, such as the situation of washrooms or cafes, to minimize the effects on the environment which might be caused by water, steam or foodstuffs (which might attract insects). It also deals with zoning within the building (eg rooms, display cases, zones in storage areas) in order to achieve appropriate environments for particular types of material (eg to provide an area with low light levels for the display of watercolours or other light-sensitive material) (Fig 20). The way in which the environment is managed is highly dependent on the type of building (eg whether it is a significant historic building which cannot be modified). There are, for example, considerable constraints in managing the environment in a historic house which is intended to look as it did in the eighteenth or nineteenth century (Staniforth *et al* 1994).

An important aspect of environmental management for museums, monuments and sites is visitor management. This may include positioning of car parks, control of visitor numbers, situation of cloakrooms so that visitors are persuaded to leave their outer clothing before entering display areas, routing of visitors through buildings to avoid congestion or overloading parts of the structure (Lloyd & Mullany 1994, Yates & Davison 1996, Périer d'Ieteren 1998).

Figure 20 Part of the stored archaeological archive resulting from recent excavations in London (Courtesy of Museum of London).

Reactive conservation

Planned conservation is possible only in the context of a wider management policy, so a collection or monument may be the subject of condition survey, or a programme of care and maintenance, or of a rolling programme of investigation and elucidation (eg the cleaning of a collection of easel paintings), but much conservation is reactive even within the framework of a management plan, in that it happens in response to a range of possible events.

Some, of course, is reactive to unplanned events such as flood or fire, but much, possibly most, conservation action is instigated by some change in the status or use of the monument, collection or individual object. An object is revealed during excavation and needs attention to enable it to survive the process of extracting it from the ground. An object is to be acquired by a museum and is assessed by a conservator prior to purchase. A decision may be made to exhibit an object or open a monument to the public for the first time, which may mean that conservation processes are needed to make the object/monument clean, structurally sound and understandable for display. Equally, if an object or collection is to be repatriated, or lent to another institution, or to become part of a teaching programme, it may need conservation assessment and possibly treatment in preparation for its new role. If an object is to be returned as far as possible to its intended function (eg a damaged painting to be made legible, an ethnographic object to be made available for use in a ceremony, a clock to be made to tell the time) conservation treatment may be needed. In these situations conservation is employed to facilitate use either by putting the object in strong enough condition, or through treatments which aid interpretation.

Although good storage or display conditions may be established for a whole collection most items will be given individual conservation attention only once in period of several years. Attention 'little and often' (ie regular minor maintenance) may be more effective, and may serve the principle of minimum intervention better than one major exercise, but because of the size of most collections (particularly archaeological or natural history collections which tend to contain very large numbers of objects), this regular attention is difficult to manage.

Major objects may get more frequent attention and this can have its dangers in that it is more difficult to adhere to the principle of minimum intervention because there is always the temptation to handle and examine a 'famous' object.

Documentation

Documentation is essential in the practice of conservation. It consists of written or computerized records of the state of the object and any conservation procedures undertaken, together with photographs, diagrams and analytical data. Clear documentation should leave no doubt now, or in the future, about what procedures have been applied to an object or collection, why they were applied and what the results were. It provides a permanent record of what has been discovered, such as technological details, during the course of conservation, and reference to this record helps the conservator or curator to understand the present state of the object.

Documentation must be accessible so that the information is readily retrievable in the future by anyone with a legitimate interest in the object or collection. Concise, well-organized and carefully illustrated documentation is far more effective than wordy and unstructured discussion (Keene 1983, Corfield 1992, Umney 1995). In order to ensure some consistency most documentation is undertaken using a structured form or check-list, and many institutions have adopted a set of standardized terms, particularly when describing condition, or frequently-used treatments. However, much of the terminology used is still too loose and unspecific to provide effective consistency. Many conservators now use computerized documentation, but an issue faced in museums and related institutions is how to integrate collections and conservation information (Keene 1996), particularly so that conservators have access to the accession data and curators have access to conservation records (Dollery & Henderson 1996). This situation should improve now that collections management embraces all aspects of objects and their care, and more museums are adopting computerized management systems.

Unfortunately, good standards of documentation have developed only in the last quarter century, and standards are still very variable, certainly between conservators working in institutions and those working for private clients; thus information on past conservation treatment of objects is still patchy and where it does exist it may not be very informative.

Uses of documentation

Documentation is important because it records evidence revealed during investigation, and other procedures applied to an object, in much the same way that records of an excavation provide data on all the layers and features uncovered. Just like excavation, much conservation involves removal of evidence (eg in 'cleaning'); thus keeping all the information is 'preservation by record' of the perceived state of the object before treatment, and all the information derived from it during the process, all of which contribute to the significance. Recording the stages of any conservation procedure which changes the object (such as removing superimposed limewash from a wall-painting or removing later upholstery from furniture), provides data on later additions, on the stages of removal or dismantlement, and on the earlier state. In the process earlier repairs or restorations may be detected which aid understanding of present condition and history of use.

Documentation of condition is well-established as an essential tool in the management of change where loans are concerned. Here, documentation enables the condition at the time of lending to be compared with condition after transport to another location, or on completion of an exhibition. Monitoring and recording changes in condition also has potential for research into effectiveness of conservation treatments. Although Stone (1996) found that evaluating old treatments was limited by the way in which records had been kept, in some institutions it is now becoming possible to use documentation in this way. At the Museum of London there are records of the conservation of over 35,000 objects and a project has already been undertaken to assess the effectiveness of glycerol treatment for waterlogged leather (Ganiaris & Starling 1996, Sully & Suenson-Taylor 1996). As the value of being able to use documentation in this way becomes more apparent, and as standards of documentation are improved and become systematized, it should become increasingly possible to research the effectiveness of past treatments (Suenson-Taylor *et al* 1999).

Keeping a record of condition helps to detect trends in deterioration and enables conservators to intervene in order to manage any change which is taking place. Tracking condition can also help to facilitate use of objects since current use can continue if there are no apparent indications of change in the object. Another way in which documentation safeguards the integrity of the object is in minimizing the need for further work on the object itself. If conservators can check what has been done in the past this minimizes the need to apply potentially damaging investigative tests to the object itself (eg to identify a conservation adhesive). Results of analysis, and samples, should be kept with the record and be available for study and possible reinterpretation, thus further testing on the object itself can be limited to cases of doubt, or where information is sought which was not looked for in the past. Documentation should provide full details of all materials used during conservation procedures since it must be possible for a conservator working on the object in the future to understand and deal with any subsequent changes to these materials, or interaction with the object, and to choose a suitable method for retreatment, if necessary.

Defining the problem

Conservation takes place in response to the recognition of a 'problem' which needs a conservation solution. It follows that the first stage in conservation is to define the problem. Conservators dealing with large quantities of material where individual items may have only moderate value (eg archaeological ceramics, or large collections of paper) are used to defining the problem rapidly and summing it up briefly in documentation, but for large or complex objects or structures, or when dealing with a whole collection as an entity, arriving at a definition of the problem will require several forms of assessment.

Assessing the state of the object

Because conservation has the power to change objects both materially and conceptually, conservators must assess the object carefully before making any decisions. To achieve a complete picture of the problem it is necessary to evaluate significance, condition, context (in the sense of both setting and environment) and use. Other types of information relevant to this assessment and to future action are the object's original function, the details of its past and present ownership and environment, and something of its history both before and since being acquired by the museum.

Sources of information

For objects in collections, or structures with well-established reputation, information on history and significance may be readily available through existing records or publications, through discussion with colleagues, or the owner, or by consulting makers or their successors. For some objects there will be no records or documentary evidence, and, where there is, these may be patchy, unreliable and influenced by the views prevailing at the time they were written. Equally consultation may not provide 'all the answers'; specialists cannot know about every type and variation of object, and owners may have very specific views related to their own valuation and use of the object. Makers or their successors may sometimes have changing or conflicting views. Conservators, of course, develop considerable knowledge and expertise concerning the types of object on which they normally work (eg ceramics, furniture, or locomotives) but cannot necessarily establish details of the 'whole' object without consultation.

Visual examination, looking at objects

One of the most important skills which conservators (and curators) develop is how to look at objects, a frequently undervalued skill since there is now a battery of scientific techniques which can be applied to objects. This kind of looking is a deliberate act and requires time, concentration and practice. Learning to look effectively at objects involves storing visual information, adding to it with every new experience and internalizing it so that it supports visual understanding.

Visual evidence is questioned and compared to what has been seen in other objects, so conservators draw on their experience and knowledge and learn where to look and what to look for.

> *We only see what we look at. To look is an act of choice. As a result of this act, what we see is brought within our reach. (Berger 1972, 8)*

This is like to listening to music. Background music is not consciously listened to; to hear detail it is necessary to pay attention to the music, shut out other distraction, and concentrate. It then becomes possible to discriminate between different instruments, to detect repeated phrases or themes, to recognize the style of a particular composer.

However experienced and however careful the conservator is, looking is still subjective and selective. Each time a conservator looks at an object s/he will probably see something 'new'. It is also important to note that looking in this way involves surfaces only, even when looking underneath, or inside a hollow object (van de Wetering 1996). On some objects the qualities of the surface can indicate something about what lies beneath (as when the underlying structure of a wooden object is mechanically weakened by insect damage), and information can be gained, for example, by looking at broken edges, but other techniques are needed to examine the underlying structure of an object in any detail.

Surface investigation

Objects obscured by accretions of dirt or salts on the surface or by layers of corrosion need to be investigated in order to discover shape, surface and surface details. This process is sometimes referred to as investigative cleaning and can be compared to an archaeological excavation or a forensic exercise in that it involves careful examination of the accretion in order to decide what is relevant to the significance of the object, and what is not (thus what may be removed to clarify significance). The aim of this process is to establish the state of the surface but not necessarily to remove all accretions; many archaeological objects are partially cleaned only, taken to a stage where type and function can be understood.

Assessing significance

Material significance

Assessing material significance is a matter of examining the different levels of structure in an object and it involves moving from the visible information on the surfaces to the 'invisible' information within the structure of the object.

> *Hidden from direct observation, but detectable by instrumental means are many characteristics that can be used to support authorship, attribution, provenance or date. Some of these characteristics relate to composition ... others relate to structure. (Meyers 1988, 5)*

Conservators look at the visible surfaces either with the naked eye or using some form of magnification such as a microscope. They will also consider factors such as weight, and they may investigate the surface, for example by minute probing or by using a fine brush to remove overlying dust in a small and inconspicuous area. Tiny samples such as an individual fibre or a delicate scraping of pigment particles may be taken from the surface for closer examination. Visual examination takes note of shape and form, colour, texture and gloss, wear and damage, loss, repair and replacement, as well the different materials present and how they are related or joined together. Objects may include unexpected materials such as old repairs and fills (Thornton 1998) or unusual inclusions (eg parchment and gelatine in textiles, see Kite 1992).

Looking at the inside or underside of objects provides further information, and using a microscope reveals surface detail such as tool marks on metals or bone, or structure of threads from textiles (Hunter & Foley 1984, Darrah 1987, Fell 1996). Other techniques may clarify surface features, such as use of ultra-violet light to differentiate between original and recent additions, or to detect damage, or infra-red photography to clarify design on a painted object (Sayre 1993).

Looking at the internal structure involves either using some form of radiography, eg X-radiography of metals (Lang & Middleton 1997), to penetrate beneath the surface (Figs 21 and 22), or taking a sample in order to look at internal features such as the cellular structure of wood (Hather 2000) or the layered structures of paintings. A highly magnified image of surfaces or structure can be obtained by using a scanning electron microscope (SEM) to examine materials such as textile fibres, structures of faience and frits, or surfaces of metal objects (Tite & Bimson 1987, Janaway & Scott 1989, Goodburn-Brown 1996) (Figs 23–25).

Chemical tests can be used on tiny areas of the surface of an object to check composition (and these can be useful for conservators who do not have access to analytical facilities) but it may be more effective to use some form of instrumental technique to detect elements or compounds present, for example X-ray diffraction (Daniels 1989, Lambert 1997). Because of the cost of the equipment few conservators have direct access to such methods and, when necessary, rely on sending samples to a materials scientist. This is where conservators must

Figure 21 A pair of Roman shoes concreted together during burial (Courtesy of Museum of London).

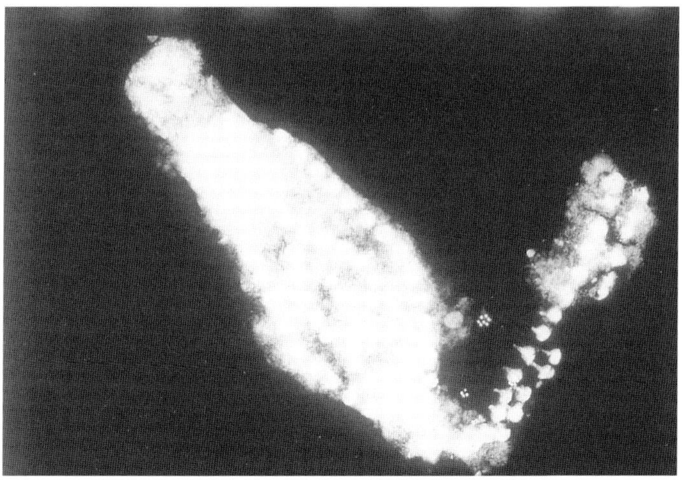

Figure 22 Investigation: X-radiography. The X-ray of the shoes revealed iron hob-nails and copper studs (Courtesy of Museum of London).

understand the potential of the techniques being used and frame their questions in such a way that the scientist can focus the analysis on relevant aspects of the structure. Like visual examination instrumental examination will detect only what it has been designed for, or set up to look for. Different techniques will detect different ranges of elements, none will detect all of them. Once the question is framed it becomes possible, for example, to examine distribution of elements in corroded glass surfaces (Schreiner 1987) or salt crystals in the Dead Sea Scrolls (Wallert 1996) or identify pigments in wall paintings (Bearat *et al* 1997).

The information derived will help to elucidate the condition of the object, and guide conservation decisions. There are good reasons for establishing the component materials of the

Figure 23 A Roman buckle with traces of wood from the associated coffin (Courtesy of Museum of London).

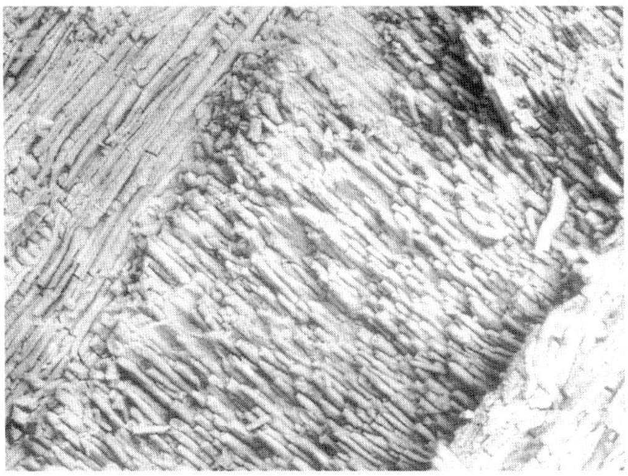

Figure 24 Investigation: Scanning Electron Microscopy (SEM). Magnified structural details of wood (probably oak) from the Roman buckle in Figure 23 (Courtesy of Museum of London).

Figure 25 Investigation: Scanning Electron Microscopy (SEM). Magnified details of tool marks on the buckle in Figure 23 (Courtesy of Museum of London).

object, and the types of manufacturing process. Conservation procedures may be suitable for some materials but not for others (eg robust bone or a well-fired ceramic can be safely cleaned in water but ivory or a sun-dried clay object could be destroyed by the same treatment), and information on manufacture such as tool-marks or traces of original adhesive could be removed or changed if their significance is not understood at the outset. Understanding of materials may aid assessment of date and place of manufacture and interpretation of possible repairs and alterations. Knowledge of materials may be relevant in other ways as in matters of health and safety (some materials, such as lead, are inherently toxic; some objects may have had toxic materials, eg pesticides, applied to them: Johnson 1999).

Elucidation of excavated objects

One of the difficulties with examining excavated archaeological objects is that their shape and surfaces may be partially or totally obscured. Thus for some deteriorated archaeological materials the very first stages of examination may involve attempting to discover what the object is (eg single object or mass of objects, whole or broken, brooch or key etc). This may mean establishing the outline of the object(s) through using X-radiography to penetrate the overlying soil or encrustation (see Figs 21, 22), or carrying out a mini-excavation through part of the superficial accretions in order to establish the type and condition of the surface of the object. During this form of examination, and during subsequent treatment of the object, much useful technological information is gradually revealed. For example, a highly corroded green metal disk may turn out to be either a coin or a button or possibly a brooch, it may be either bronze or brass or predominantly silver. If it is a coin it may be of almost any date (after the introduction of coinage) and establishing this information depends on the conservator's ability to elucidate the information (Casey & Cronyn 1980, Cronyn 1990).

Detailed investigation of the object can establish what substances are present and how they have been used. This enables confirmation of the material-type (eg ceramic or stone), and facilitates understanding of technological achievement in the form of selection and manipulation of raw materials (such as methods and materials used to make pots, or type of wood used to make a shield). It can also confirm similarities of technology or raw materials used by different groups; or presence of rare or foreign materials (eg elephant ivory, silk).

This technological investigation is crucial in order to understand the object, but some of the information revealed may be more relevant to the archaeology of the material than its conservation. The details which reveal the type or date of an object may have no bearing on the choice of treatment but may be crucial to the interpretation of the site. Much of this type of work is a form of archaeological investigation. Thus conservators may reveal the construction of objects such as an early medieval reliquary made of wood encased in decorated metal sheet (Keene 1987), or traces of different organic materials preserved in association with a corroded sword, shield and other grave goods, which provide evidence of the existence of a scabbard, and a textile bag (Watkins & Cameron 1987).

Assessing conceptual significance

Much of what is bound up with conceptual significance cannot be seen on the surface of the object, nor can it be detected by looking at the materials present in the body (though some material characteristics may be indicators of how the maker or user viewed the object). Many aspects of conceptual significance are hidden in the sense that they are apparent only to those who are aware of the role of the object in, say, secret or sacred ceremony, or who have studied the artist's intent. Faced with treating a collection of Masonic aprons Lord found that:

> *scant information was available on the aprons and little has been published on them so it was necessary to research their background. Also, the subject of Freemasonry was unfamiliar, making it difficult to know whether to approach them as ritualistic, historical documents, or art objects. (Lord 1996, 95)*

Assessing conceptual significance is essentially a matter of consultation and discussion though it will certainly start with visual examination and this examination may be a collaborative exercise. Conservators need skills of questioning, sensitivity to other views and the ability to work with other people in order to establish the meaning of the object. Thus Reedy described consulting Tibetan priests in order to inform the conservation treatment of Tibetan religious art (Reedy 1992) and Clavir provided records of how First Peoples in Canada experienced the meaning of objects (Clavir 1994, Clavir 1999). Among others, she quoted the words of Pete Jemison, 'when you get to the level of sacred objects, they shouldn't even be in collections with a curator. They should be back among the people who handle and care for them' (Clavir 1994, 53).

Conservators aim to discover how the object is valued and by whom, and what it can indicate about the people who made and used it. There may be a number of different people with views on its significance and conservators need to take this range of views into account. Curators will be concerned with the role of the object within the collection and the way its meaning can be interpreted to visitors, contemporary artists have views about the intent of their works and how they should be dealt with in museums, indigenous peoples may wish to re-establish an object's traditional function which, again, will affect the way in which it can be handled or displayed.

Understanding the artist's intent is important when dealing with art objects and can be particularly challenging when dealing with contemporary art. Much contemporary art incorporates 'strange' materials or experimental technologies but the conservator needs to establish whether the artist intended the material to change or even disintegrate, or whether s/he hoped the art would maintain its original appearance. Establishing the artist's intent is seen as important when dealing with sculpture which is exhibited in the open air and affected by handling and weathering. Conservators have attempted to establish the views of artists such as Henry Moore or Barbara Hepworth on how they would have preferred their works to be conserved (Pullen & Deighton 1995, Summers 1995). Baldwin (1999) writes of Robert Rauschenberg that he accepted the signs of change in his art works and did not want them changed by restoration, and Magor (1990) worked closely with the conservator in deciding how best to maintain the conceptual significance of one of her works after some of the bottled fruit and vegetables which it contained started to decay.

Conceptual significance of archaeological materials

Conservators work with the idea of conceptual significance as applied to ethnographic artefacts and to art works (objects of the 'other'), but consider it less frequently in relation to archaeological objects (objects of the 'past') except, perhaps, to classical art. Archaeological objects have been 'sanitized' by the fact that it is possible to 'know' little about their meanings, or the intent of the maker. For an excavated object the context in which it is found may indicate its intended use, eg a pot encrusted in sooty deposits and lying by a hearth may be a cooking pot, a diadem in a grave may have been a prized personal possession, or even a symbol of status. However, it is the object itself which is the main source of information, and many excavated objects appear to have been carefully shaped and ornamented to give them visual impact, so may have been selected and prized by their owners for their aesthetic qualities. Thus at least some of the objects found in graves probably had conceptual significance for their owners, and at least some archaeological objects must have been intended by the maker to have a specific appearance or use (maker's or artist's intent).

Archaeological conservators have been trained to take a scientific attitude to excavated objects and tend to concentrate exclusively on material significance and the object's role as evidence, but they should be open to the aesthetic qualities and other possible intangible meanings of archaeological objects (ARAAFU 1996). Some ancient objects could certainly be considered to be works of art, and were made with great skill and delicacy, such as a fourth-century BC Macedonian gold wreath (in the J Paul Getty Museum collection) which imitates myrtle leaves and flowers with botanical accuracy. Maish (1995) writes about the importance of considering both the aesthetic and the evidential character of this object while conserving it, and this is an example that other conservators should follow even when they are conserving less obviously beautiful objects. Conservators and curators should be aware that conscious or unconscious views about, for example, 'crude' prehistoric pottery or 'beautiful' classical Greek ceramics, affect their practice (Bertholon 1996).

Furthermore, to consideration of original significance and artist's or maker's intent, should be added the conceptual significance and aesthetic qualities of excavated objects for audiences today, who feel that patina and other signs of age are crucial to their appreciation of archaeological objects. Disregarding this aspect has led to insufficient thought being given to the visual impact of conservation processes, resulting in objects robbed of the magic of age – just as some excavated sites lose the charm of romantic ruins once they are tidied up and neatly presented, 'the moment man steps back into the scene, as restorer, preserver and even as visitor, the charm is gone' (Morin 1999, 193). Ancient silver objects which are cleaned and polished can look new and ordinary, carefully tended historic houses can look as if they were never lived in.

Assessment of condition

Assessing condition, whether it is of a single object or a collection, or is part of a regular monitoring procedure, is essential in planning conservation because the condition of the object affects the choice of procedure. Condition also affects the significance and use of an object and in this way may affect conservation decisions (Ashley-Smith 1999a). Condition is a relative property and the conservator should consult earlier records (if any) and assess the condition in relation to the object's history and context (so assessment may also involve assessment of the environment in which the object or collection is stored or displayed). The assessment results in a description of condition (or condition report) and here some standardization of terms is essential to attempt consistency (Buck 1971, Keene 1991, Keene 1996). However, an acknowledged weakness of condition assessment is that it is subjective and depends on the individual conservator's experience and view-point. One conservator's reports may be internally consistent but it may be difficult to achieve consistency within a team, or to find consistency when using earlier reports undertaken by others. Newey *et al* (1993) compared the way in which conservators from two different institutions assessed iron objects: there was consistency within each institution but not between them.

Condition of individual objects

Assessing condition should involve the conservator in working out what changes have taken, or are taking, place in the object or structure, and how much of a threat they are to the stability and legibility of the object itself. It requires the conservator to ask a series of questions while examining the object (these may be prompted by a standard checklist), and not to rely purely on what is visible, but to consider underlying features and to use further investigative techniques such as X-radiography, where necessary.

The condition of the object as a whole should be considered, as well as the condition of the component parts or materials. The object as a whole may be in quite good condition but one feature or material may be in very poor condition (eg the hinge which attaches the lid to a box, or the enamel on a copper-alloy object in otherwise excellent condition). The object may be in an unexpectedly stable condition considering its history and context and this may guide the conservator towards a comparatively minor intervention, whereas another object or structure may be in exceptionally poor or even parlous condition in spite of apparently suitable care measures and here the conservator will be forced to reassess the situation and perhaps apply a more thoroughly remedial treatment.

Earlier treatments and condition

Assessment should involve looking for records, or indications on the object of earlier conservation procedures. These need to be differentiated from original repairs which are part of the history of the object, and tests may be needed to establish what kinds of materials (eg adhesives) have been used. Original repairs or conservation treatments may be making an important contribution to current condition. A successful treatment may be left alone, or reinforced with a similar approach; an unsuccessful one or a repair or treatment which is unsightly or causing further damage, may be, as far as possible, removed. Sometimes materials introduced into the object during earlier procedures may not be readily removable and may leave residues which may affect the object in the future (Daniels 1999). Some treatments may have exacerbated deterioration, eg by depositing salts in ceramics (Paterakis 1996). On the other hand, some early interventions (such as the use of metal rivets to repair ceramic objects) are interesting in their own right and may now be considered part of the history of the object and worth retaining, provided they are not causing any serious damage (Koob 1998, Thornton 1998).

Monitoring condition

Assessment of condition in the form of monitoring condition is a useful tool in managing change. Condition should be checked periodically during a conservation project, and conservation procedures adjusted accordingly. The point of supervising loans and transport of objects from one destination to another is not just to prevent loss but to monitor and record condition, so that a loaned object can be withdrawn if, say, the stresses of a lengthy travelling exhibition are causing unacceptable change (and so that, in the event of dispute over causes or effects of damage, there is a good record of condition beforehand). Monitoring condition of recently excavated material while stored on site can make an important contribution to the care of this material. Monitoring condition of permanent collections at regular intervals can enable conservators to assess the effectiveness of environmental management and preventive care, including, for example, the effects of handling or the pollutant effects of materials used to construct display cases.

Condition surveys

In 1988 a report by the National Audit Office on the UK national collections revealed that not enough was known about the condition of museum collections (Sheldon 1988). Since 1988 many museums have carried out condition assessments or surveys of their collections and have used the results to develop policy for collections care and management (equally, condition surveys may be carried out on monuments and sites as part of the development of conservation plans, and in storage areas on an archaeological site). The purpose and scope of condition surveys must be clearly defined at the outset so that the relevant type and quantity of data is obtained to enable the objectives of the survey to be met. For example, the data required to assess what proportion of a collection needs remedial conservation, and the resources required for the work, will differ significantly from that required to identify causes of and trends in deterioration, and effectiveness of earlier conservation treatments (Keene 1996). Assessment of a collection may involve examining every object, but where large numbers of objects are concerned the collection may be sampled by, for example, examining every tenth object and using a statistical approach when interpreting the results. The most important result of a condition survey is that conservators can plan their work from a position of improved knowledge of the state of the collection (Oakley 1990, Keene 1991, Buisje 1995, Keene 1996, Moore 1996, Suenson-Taylor *et al* 1999, Talley 1999).

Assessing context

Context, in the sense of setting, can influence conceptual significance and can thus affect options for conservation. Archaeological objects derive much of their significance from the context in which they are found and the materials they are associated with, and (by association) they can affect the significance of other artefacts. Thus destruction of relationships to the assemblage and to the location during illicit digging results in loss of information. Conservators working on site may excavate a group of fragile artefacts together in order to preserve the assemblage for detailed study; investigation and conservation of assemblages (rather than isolated objects) enables conservators and archaeologists to gain maximum benefit from the material.

Some artefacts such as architectural features or furniture have been designed and made for their setting and should be preserved *in situ* if at all possible (Philippot 1976). Drysdale and Rowe (1998) discussed the return of two ornate gilded tables to their original context in Chiswick House, London, and the choice of conservation procedures used to reinstate their significance in their old context. Perhaps the most extreme example of loss of value caused by artefacts being divorced from their original setting is the removal of wall paintings for display elsewhere (Moormann 1991), although sometimes, of course, this is unavoidable. The paintings lose both material integrity and conceptual significance, and attempting to replace them in their original positions in order to regain conceptual significance is not necessarily successful (Mora *et al* 1984, Brajer 1999).

Almost all museum objects have been removed from their original contexts and given a new context in the museum, but in some instances the museum context itself has become significant as, for example, in small private museums established within historic houses, or museums with a specific intent and character such as the Pitt-Rivers Museum in Oxford, UK (Blackwood 1970). In these situations the historical significance of the context, such as how the objects are grouped and displayed, and the cases they are displayed in, should be preserved (even if they do not accord with modern conservation standards). Conservators must be not only sensitive to context, but plan their conservation measures to safeguard both the individual objects and the context. Thus the character of a display will be preserved although means may be found to improve conditions for the objects unobtrusively, eg by controlling the climate in the cases and by providing internal supports.

Context, in the sense of environment, affects condition and thus both material significance and the type of conservation procedure that can be applied. For example, decorated window glass which is still *in situ* in a building may be highly corroded and may need to be cleaned and protected from the effects of the weather, but, given its setting, it must also allow light to pass through the glass so that the image remains legible. Windows which have been removed for display in a museum (and thus to a different environment) are not exposed to such harsh conditions, so may not only be in better state but can be treated in a different way (Newton & Davison 1989). Preservation *in situ* provides particular challenges, as in the treatment of the tiled floor and wooden sill beams of the Archbishop's Mint at Trondheim, Norway (Peacock & Turner-Walker 1998), or the difficulties of deciding on appropriate conservation measures for Queen Nefertari's tomb at Thebes, Egypt, which has suffered from intractable environmental problems, including salts (Preusser 1991).

Assessing use

The way an object or monument is used is linked both to its significance and its condition (Ashley-Smith 1999a). Its current use may be responsible for its poor condition, as with heavily used books or documents which are handled regularly, thus assessing the condition of an object should involve examining its current use. Potential or future use has an important impact on choice of conservation procedure, and can be assessed and agreed only through consultation and discussion. Although some forms of use may appear to be in opposition to conservation, conservators should bear in mind Ashley-Smith's statement that 'the only unethical treatments are those that limit future use' (Ashley-Smith 1999b, 132).

Some uses will require much more extensive remedial measures than might otherwise be needed, eg a coin or medal required for exhibition may have accretions removed from the surface in order to make it understandable, a steam engine which is to be in working order may have some of its working parts replaced so that the whole mechanism functions smoothly and does not put other original parts at risk. Conservators need to be aware of the wider context in which they work and some are now publishing case studies which illustrate how considerations of future role and use of an object affect conservation decisions. Good examples include Smith (1994) on conservation and restoration of ceramics, Eastop (1998) on textiles and Cruickshank *et al* (1999) on painted Egyptian shrouds. The considerable constraints placed on conservation by the use of certain types of artefact are very well illustrated by Pavelka who discussed the need to keep archival and library material in usable state, and the consequent compromises required in conservation treatment (Pavelka 1999), and by Farnsworth who discussed the situation of religious sculpture which was still 'in active service' and must be maintained in a state fit to be carried in procession or used in other ceremonies (Farnsworth 1997).

Limitations of object assessment

All examination from simple visual techniques to sophisticated instrumental analysis is selective. For example, infra-red analytical techniques applied to paintings can reveal underdrawing executed in charcoal but not in organic pigments (Bomford 1998), and the higher the lead content of an alloy, the less effective is X-radiography. Some materials are easier to analyse than others; for example pigments have been successfully analysed (using chemical and, later, instrumental techniques) for two centuries, while only relatively recently has systematic study of media been possible (and this is still limited to relatively few centres) (Rees-Jones 1990, Bomford 1998). Many techniques require a small sample to be taken from an inconspicuous place, but a single sample may not be typical of the whole and only what is present in that one spot will be detected. In order to choose appropriate methods of examination and analysis (and to take suitable samples) the conservator, archaeologist or scientist must know what s/he is looking for. Furthermore, the results of all analyses require interpretation: degraded or mixed samples may be particularly difficult to understand, and thus could be misinterpreted.

No form of examination can reveal 'all there is to know' about an object, but comparing the results of several techniques and taking account of different views will, together, provide a reasonable interpretation of the conceptual and material aspects of an object, within the present state of knowledge. There is almost certainly information within objects which cannot be tapped with present techniques, and objects should be left in a state which will allow future curators, conservators and scientists to discover more and evolve their own interpretation.

> *The almost limitless information stored in any artefact offers powerful incentive for conservation. Yet the reality is that in one lifetime (or the object's one visit to a conservation facility), we shall reveal or capture but a minute fraction of that information, and even then we may have to make hard decisions about what information is of most interest and significance ... Real limitations on resources restrict scientific investigation to a minute sample of artefacts. Hence the need for perpetual care, so that later generations may address different questions to the object, may discourse with the various levels of information it contains, in the light of their perceptions and preoccupations. (Leigh 1994, 273–4)*

Looking as an act of interpretation

The ability to look at objects in a critical way and to draw on knowledge of materials and objects is a crucial skill and when applied by a specialist to a work of art it is sometimes known as connoisseurship. Precisely the same skills of looking and evaluating can be applied to any object. Some hold the view that this is essentially an accurate method of assessment.

> *That term [connoisseurship] is unfortunately maligned because it seems to smack of preciousness and elitism. But in fact connoisseurship is a powerful scholarly tool permitting rapid distinction between what is true and what is false. I know of no other field of historical inquiry in which it is possible to achieve such rapid and precise analytical results as connoisseurship, which enables an analyst to say immediately and with assurance that this chair is Philadelphia, 1760–70 ... or this chest is Essex County, Massachusetts ... about 1670. (Prown 1993, 4)*

However, it is not only subjective but it is also a skill of interpretation and is subject to fashion, current views and knowledge, so it cannot be said to reveal 'truths'. It is well-known, for example, that the work of connoisseurs of paintings is frequently re-evaluated as attributions to specific artists are reviewed and changed (Phillips 1997). In this sense it is possible to agree with Prown that it is a powerful tool but not with his view that it is exact and unerring. It can be very effective when used by an experienced exponent, but should never preclude the use of other techniques, such as scientific analysis, to test conclusions drawn from visual examination.

Reaching a decision

Reaching a decision on whether to undertake any conservation, and if so which procedures to use, involves firstly defining the problem and then deciding what the objectives are for dealing with it. So, for example, if an object is to play an important part in a display but has been found to be both in poor condition and difficult to interpret, a decision must be reached on how to deal with these problems. Reaching a decision involves defining the objectives, identifying the possible options for conservation, then evaluating these options in the light of constraints such as available resources, and in the light of the risks and the benefits each option might bring. Some options will be ruled out because of constraints (they may be too expensive considering the available funds) or because the risks outweigh the benefits.

As with the assessment of an object, a conservator dealing with a number of similar, relatively unimportant items may deal with this process of decision-making very rapidly and without the need for consultation. However, complex situations may require consultation or teamwork in order to arrive at an agreed decision; thus much of the work on developing guidelines for decision-making has been done in the area of sites and monuments conservation (see Creigh-Tyte *et al* 2000).

Objectives

Conservators and stakeholders (eg curators) should be clear about the objectives for any conservation procedure or project, large or small, since they will guide decisions taken about conservation options. The objectives depend on the condition and on the use (and significance) of the object or structure. Condition may dictate that the objectives are either preventive, remedial or to aid interpretation. If the use of the object focuses on its research potential the objectives of conservation may be to put it in a suitable container to allow ease of examination while minimizing handling. If the same object is to be displayed then the objective may be to remove accretions, possibly to deal with areas of loss, and to provide an unobtrusive support which will allow it to be viewed in a display case. The approach to conservation of a stored collection would be very different if the objectives were to open the store to visitors, rather than to continue to hold material as a reserve collection.

Constraints

There may be a number of different constraints affecting the choice of option. An obvious constraint is provided by the ethical principles set out in the various charters or codes of practice, drawn up to guide practitioners when facing issues which arise in practice. Most conservators will be familiar with what each relevant code contains and will only deliberately consult them when faced with an unusual problem. In general, the effect of codes is that conservators should ensure that there is an agreed and valid reason for the conservation options chosen, that remedial measures (and restoration) are used only if preventive measures will not fulfil requirements for maintaining or demonstrating significance, and that any option chosen must result in a recognizable benefit. They should work only within their own expertise, and should aim to choose the least intrusive of the possible procedures which will at the same time safeguard significance. They should choose conservation materials which have well-established properties and which are compatible with the object. Finally, they should document fully all the work they have undertaken.

Another constraint may be provided by the policies of the museum, for example whether sampling is allowed for analysis, or the extent to which restoration is likely to be approved. Still further constraints may arise from the opinions of stakeholders or interest groups, for example the views of indigenous peoples about how their material cultures should be handled or conserved. Contemporary artists, too, may have strong views on how their works should be treated and here an additional constraint on conservation action may be the legal protection accorded to contemporary works of art through copyright (Curnoe 1990, Stone 1990, Fry R 1997).

There may be constraints of health and safety, time, facilities, expertise, or cost. What can be achieved in a well-equipped and well-funded laboratory is probably different from what is possible in a temporary field laboratory; what can be done within a team of conservators each specializing in a particular type of material may be very different from what can be managed by a single 'generalist'. Conservators working independently are particularly constrained by the facilities available to them and by the agreed terms and conditions of the project, including costs and time. The nature of the object itself may provide further constraints; setting aside artist's intent, some materials are difficult or impossible to conserve effectively such as ethnographic sugar sculptures and chocolate used in contemporary art (Wharton *et al* 1995, Daniels & Lohneis 1997).

Options for conservation

Not only are the problems presented by an object or collection likely to be multifaceted, but the range of conservation options which could be used to tackle them is varied too. A range of possible options will be considered, encompassing different levels of conservation from minimal intervention to more major treatments, and/or different approaches or materials used in the same type of treatment (eg different adhesives which might be used in a repair).

Broadly, the options for conservation are likely to fall into the following general categories (which are discussed in the next chapter): doing nothing; changing or modifying the use; modifying the environment; adding materials to the object, or removing materials, in order to achieve improved legibility or stability; modifying the shape, or sometimes changing the chemical structure in order to regain something of the original appearance or use. In most cases, more than one of these approaches to conservation may be needed, and for each approach there may be more than one possible option.

The implications of different conservation procedures should be considered in relation to the risks of affecting both conceptual and material significance. Thus opening a consecrated Tibetan bronze to examine and clean the contents may be avoided because of its effect on conceptual significance (Reedy 1991). Certain types of treatment will be avoided because they may damage materials present in the object (heat will not be used on glass for fear of inducing crizzling of the surface, coloured textiles will not be washed without testing all the dyes).

Risk assessment

An essential aspect of reviewing options therefore involves evaluating the risks and benefits of the various possible actions applied to objects. Risks produce loss of some of the values attributed to the objects such as use-value, or conceptual significance; benefits may include benefits to the object itself or to the museum or to the wider community. The risks to which an object is exposed may arise from the environment in which it is housed or from any action such as lending it to another institution, or applying a conservation treatment. The benefits to the object itself may be either (or both) material and conceptual. The construction and surface characteristics may be clearer after removal of accretions; the meaning of an object may be enhanced by restoring missing parts or by replacing it in its original setting. Wider benefits may include, for example, the fact that an object is more accessible to visitors if displayed or used in a certain way, or if lent to another institution.

Broadly, assessing risks involves looking at the possible disadvantages of particular courses of action. It involves estimating whether a conservation option will introduce an appreciable risk and, if so, how serious this risk is likely to be, how often it is likely to occur, and what are its possible long-term effects. Since few, if any, conservation activities are without risk, it is also important to compare the risks inherent in different options. Risk assessment in conservation has been increasingly recognized as a crucial aspect of decision-making, and has become an important tool. It has been extensively discussed by Ashley-Smith (1999a).

It is the responsibility of the conservator to recognize the needs for keeping the object in use, and to work with the curator to accept as well as to minimize the risks which are an inevitable consequence. Use in the museum sense clearly produces risks and these risks have to be assessed by curator and conservator as part of the process of balancing the effects of use against the prospects for long-term survival of the object. Some difference of view, between curator and conservator may not be a bad thing since it will ensure that both risks and benefits are explored and assessed, but conservators must not become 'nannies'. Over-protection will not benefit the object in the long run and can be seen as refusing to accept professional responsibilities towards owner, colleagues, or institution.

> *It is important for conservators to remember not to 'cry wolf'. If wolf is cried too often, the conservator's credibility is ultimately lost and the painting [or object] may suffer. A conservator, by being too idealistic, runs the risk of opting out of true responsibility for the painting. (Barclay 1990, 26)*

In other words the conservator must be prepared to work with risk, which is inevitable if objects are to be used. Since some risk is inherent in what conservators do, risk cannot be evaluated effectively without also looking at the benefits arising from a particular course of action.

When a request for a loan is refused it is because the risks to the object are agreed to be serious enough to outweigh the advantages of its being used in an exhibition in another institution. However, when an object is lent to schools the benefits in terms of education or enjoyment are seen to outweigh the risk that it may be damaged. When a painting or textile is displayed its accessibility is seen to be more important than the risk inherent in exposing it to light.

> *We place an object at risk because we judge that the value to the present of so doing is greater than its potential value to the future; an example is the use of original objects in handling collections. Museum people have to make these choices every day. (Hebditch 1998, 10)*

Risks involved in conservation decisions have to be considered, and here the Sutton Hoo helmet is a good example. Although it was well known that the first reconstruction was incorrect, for some time it was considered too risky to attempt to dismantle the reconstruction; eventually it was considered worth taking the risk in order to achieve the benefits which would result from a better-informed restoration (Williams 1992) (see Figs 46 and 47).

Cost-benefit analysis

Many decisions in museums, just as elsewhere, may be made on the basis of financial cost even though many other factors could (or should) be taken into account. Cost-benefit appraisal is a tool which has been developed to enable decision-makers to evaluate both the costs and the benefits to be gained from different options, and it has now been applied to conservation. The types of benefit which might be considered include long-term conservation, improved access to the public, meeting a particular museum objective, enhancing knowledge and understanding of the object. This type of exercise is normally applied to large projects and the analysis is undertaken by a team of people (eg curator, conservator, collections manager), who together consider the range of possible options and their likely benefits. Using this analysis can show that in some cases greater benefit may be achieved through increased cost, and can demonstrate that a wide range of opinions has been considered, thus giving the final decision greater confidence (Cassar 1998, Arthur 1999). Some other forms of cost evaluation are given in Creigh-Tyte *et al* (2000).

Systematizing decision-making

As there is a wide range of different, sometimes conflicting, factors to consider during the process of assessment of objects and decision-making, systematizing the approach by using check-lists or computer programs may be helpful (Michalski 1994, Marcon 1997, Ashley-Smith 1999a). For example, it is usual to use some form of check-list when assessing condition, particularly when dealing with complex structures such as paintings, or when surveying a whole collection. Risk assessment and cost-benefit analysis can also be handled in a systematic way (see Cassar 1998).

Some conservators have started to consider the use of computerized 'expert systems' to help to evaluate relevant factors. A system of this kind normally prompts the entry of information about, say, the nature and condition of the object and may work through a series of questions about treatment options, thus helping the conservator to reach the same conclusion an 'expert' would. Systems have been considered or developed for archive conservation, for the selection of the appropriate grade of consolidant and solution strength in the treatment of waterlogged wood, and for predicting the behaviour of salts in stone in order to maintain conditions which should minimize salt damage (Organ 1989, Cook & Grattan 1990, Curtin 1990, Price 2000b). Use of this kind of system is not well-established in conservation, and some existing systems are acknowledged to have their problems, but further research may yield more approaches of this kind in the future.

The conservation proposal

The assessment concludes with the drawing up of a treatment proposal or project plan. This will take the significance and condition of the object as a starting point and set out the rationale for treatment. It will outline the proposed treatment, including procedures for enhancing the interpretation and use, procedures to deal with existing deterioration and to prevent or slow future change, and procedures for evaluating the results of the process as well as for long-term maintenance or care. For a large or complex object or monument the proposal may require detailed consultation and may result in a carefully argued document; for less significant and more routine objects the conservator may simply make a brief note of the intended procedure(s).

Drawing up a proposal is essential where an independent conservator is to be employed to deal with a specific project. The assessment and proposal may form the basis of the brief for a job or, more formally, for seeking competitive tenders from more than one conservator (see Conservation Forum 1998, Winsor 2000). However formal or informal, adequate assessment is essential prior to undertaking any type of conservation procedure.

7 Issues in practice: conservation procedures

Introduction

In reaching a decision about the procedures to be used the conservator will have considered a range of possible options and drawn up a conservation proposal. Discussion in the previous chapter showed that conservation choices depend on a range of factors, not just the condition of the object, thus there are no 'standard' approaches. Different options have different effects on the significance of the objects, and there are no clear dividing lines between various conservation procedures since some will have more than one effect. The effect will vary too, depending on the level of conservation applied. This chapter explores, in general terms, the range of possible options.

Conservation in practice

Conservation and change

Conservation processes are intended to change the object, or its environment, and to lead to the object's long-term survival by managing the rate of change. Conservation procedures are undertaken in order to maintain an acceptable state or to improve on an unacceptable state. Unfortunately some conservation actions may bring about further unintended reactions. Conservation, therefore, is not static but dynamic, and a single procedure cannot be assumed to have solved the problem of continuing change within the object; monitoring and further action will be needed in the future (eg Ward 1986, Plowden & Halahan 1987, Oddy 1992b, Thompson 1992).

Some conservation procedures, such as 'cleaning', repairing and re-leading a decorated glass window are complex and lead to changes in the longevity, strength and appearance of the object and in its material and conceptual significance. Some conservation procedures are minimal, such as providing better packaging, but nevertheless may slow deterioration. Most preventive and remedial procedures will prolong the life-span of the object to a greater or lesser extent. Some will clarify the appearance and readability of the object (eg through reconstruction), some will strengthen and make objects safe to handle, and some, through revealing technological information or aesthetic detail, will help to establish a clearer understanding of the object and its significance. However, changes which may superficially improve the appearance of the object (it may look 'cleaner'), may have been brought about at the cost of removal of surface accretions (which may be protective or contain evidence) or at the cost of change to the internal structure (eg stress and microscopic damage to individual fibres when washing a textile).

Some superficial changes may be seen as improvements by some, but not by others. A consolidant or surface coating may not only strengthen or protect the underlying structure but intensify colour and so be seen by some as refreshing the look of an object, but by others as masking the appearance which is the result of change and deterioration. Attitudes vary. For example, the lacquer on scientific instruments may be renewed because it reinstates the presumed original appearance (Brenni 1999) but a consolidant on a matt and powdery ethnographic paint could be considered unacceptable because it changes the appearance from matt and opaque to glossy and translucent, quite unlike its appearance 'as found' (Hansen *et al* 1993).

Different approaches to conservation

Different adjectives have been used to qualify types of conservation and they can be divided into those that describe intent and those that describe the degree of interaction with the object. Thus investigative, preventive and remedial conservation (as well as restoration) all express the aim of the process, whereas active, passive and interventive conservation are normally taken to describe the way in which the conservator interacts with the object. Here passive conservation is considered to involve changing the environment surrounding the object rather than doing anything directly to the object itself, but active and interventive conservation imply working directly on the object and probably changing it.

Preventive conservation has taken on more than one meaning. The term is used to describe the intention of preventing deterioration, but it is also often considered to be synonymous with what has been called passive conservation, involving modification and control of the environment alone. This ignores the fact that many treatments, such as neutralization of acids in paper, are applied directly to an object with the aim of preventing further change. One of the characteristics which has moved conservation from craft-restoration to a scientific discipline is the anticipation and management of future change. Thus even if the initial intent of a treatment is to remedy past damage this should be done in such a way as to minimize the likelihood of future change, or it should be coupled with preventive measures. In an earlier chapter it was proposed that the activities involved in preventive conservation need clearer definition, and the terms 'preventive care' and 'preventive treatment' were suggested.

There has been an assumption that preventive measures are 'better' because they interfere less (or not at all) with the material significance (integrity) than interventive ones, but this is to assume firstly that preventive measures are limited to modifying the environment, and secondly that preventive measures make no change to significance. With a growing understanding of the meanings encompassed by or implied by significance, it becomes clear that it is not possible to state that preventive care has no effect on significance, nor that all interventive treatments automatically affect significance. It cannot be assumed that a procedure limited to changing the object's environment will not affect significance, nor that any work undertaken directly on an object will automatically change its significance. For example, if the changed environment involves a change of context (eg enclosing a state bed in a protective glass case in the setting of a historic house where all other furniture remains in the open room, or taking outdoor statuary from its garden setting into a sheltered display area), it can have a marked effect on conceptual significance. Some measures that undoubtedly prevent further deterioration (preventive treatments) at the same time can involve change to the object's significance (eg the removal of salts or soils which may be highly relevant to the burial history of the object, or to its use). By contrast, some other forms of preventive treatment, such as removal of superficial museum dust, or of recent mould spores, or even of some forms of current repair, may have minimal effect on significance.

When making choices of treatment, consideration of the degree to which different forms of conservation are likely to affect significance and use, ie consideration of the outcome of the procedure in terms of significance, is more helpful than the present, rather blurred, distinctions made between intent and degree of intervention. The examples given above are, of course, debateable but the injunctions that conservators should preserve the integrity of objects and adopt minimum intervention as a guiding principle are intended to limit the possibilities of changing significance.

The continuum of conservation

The processes of conservation are not always distinct one from another in aim or effect (Brooks *et al* 1994, Berducou 1996). Although conservators talk about apparently separate activities such as 'preventive conservation' or 'reconstruction', the processes can be seen as features of a continuum from initial investigation through to restoration, and many serve more than one purpose. The initial investigative phase which may involve removing accretions from an excavated object not only reveals material information (establishing the nature of the object) but may remove damaging materials (preventive treatment), and, if taken further, may restore the shape and something of the appearance (restoration). Repairing a broken ceramic or stitching a fragile textile to a support (remedial treatment) can also be seen to contribute to restoration. Removal of damaging salts from within porous ceramics and stone will prevent future damage, but will also aid successful remedial treatment of the crumbling surface: unless the salts are removed it is unlikely that attempts to strengthen the object will be effective. Building a sympathetic internal support for a costume, normally considered an aspect of preventive care because the support will distribute the weight of the costume while it is on display, will also restore the shape of the costume so that the viewer can understand better what it would have looked like when worn.

The treatment of waterlogged wood is a good example. Keeping it wet in the field prevents collapse and distortion (preventive care); treatment with a polymer which bonds to the damaged cellulose prevents disintegration (preventive treatment) and at the same time remedies the effect of deterioration. After treatment individual pieces of wood may be joined to reconstruct the shape of the object (remedial and restoration) and the object will be given continual preventive care (Pearson 1987).

Not only is there this continuum, or overlapping, of conservation processes but there is also the paradox that conservation can destroy. Thus removal of 'dirt', for example in washing a textile, can 'improve' the appearance (restoration) and minimize some forms of future damage, for example by removing acids or mould spores (preventive treatment). But looked at under a high-powered microscope there may be signs that the swelling of the fibres which occurred during washing has caused a permanent change (damage) (Cooke 1990). Consolidating a porous surface weakened by salts without either removing the salts or ensuring adequate penetration of the consolidant may cause damage to the surface and all the information it carries (Giusti 1999). Cleaning away accretions on a coin is undertaken to clarify the image and make the coin identifiable, but it also has the power (even inadvertently) to distort details such as the mint mark and so change the significance and archaeological value of the coin (Casey & Cronyn 1980).

Thus not only do deteriorating objects present multifaceted problems, but conservation actions will inevitably be variable in their intention and effects. Taken together, these factors require that conservators consider very carefully both the object itself and all the possible consequences of the proposed conservation actions (as far as it is possible to predict them).

Selection and levels of conservation

Early versions of some codes of practice enjoined conservators not to discriminate between objects, to give the same care to every object (Sease 1998), but it is not possible to conserve everything to the same level, and this is now recognized. Conservation can never be absolute and conservators apply different levels of conservation to different objects.

There are several reasons for different objects being given different levels of attention. The amount of assessment and conservation any one object is given is, of course, influenced by how significant (or how 'useful') it is perceived to be at the time. Major national monuments cannot be allowed to crumble because there would be a public outcry, so normally they are well-maintained: in contrast minor monuments may get inadequate attention. Changes in fashion and thus in significance may mean that hitherto disregarded classes of object or building may become the subject of conservation projects (and major objects or monuments may suffer from too many attempts to conserve them).

Archaeological material often needs elucidation in order to establish how informative (how useful) it is, so anything which looks potentially informative may be investigated further. Some materials are easier to conserve than others, for example conservators have more effective means for conserving archaeological bronzes than archaeological iron, and bronzes are often more obviously interesting (perhaps decorated) than highly corroded and misshapen iron, so iron may get little attention. Some materials are very difficult to deal with, for instance deteriorating early plastic objects – other than permanent storage in a deep freezer, there is as yet no apparent way to retain their original characteristics (Reilly 1991, Fenn 1995).

There are examples of scales of treatment devised to match either condition or value of objects. This concept lies behind the UKIC Archaeology Section's proposal of five levels of conservation to deal with the enormous range of material coming from excavations (UKIC Archaeology Section 1982). A similar designation of specific levels of care for different categories or values of object was made during the Delta Plan (Ashley-Smith 1999a, Talley 1999). Using this general approach, all museum objects, or monuments, should be given the basic care of assessment and preventive care or maintenance. Both objects and structures may be given some minimal cleaning in order to clarify their nature and condition. In some cases, the condition, the context, the significance, or the proposed use of the object may dictate a more detailed approach involving more extensive investigation, remedial conservation, and possibly some restoration in order to prepare the object or structure for display. In other words, at one end of the spectrum there is preventive care and minimum remedial work, and at the other end there are 'the full works', including restoration.

Current limitations

At its best, conservation enhances the use and understanding of objects and structures, but it also has its limitations. Change is difficult to prevent and conservation cannot achieve stability for all objects and structures. Some materials are inherently unstable and deteriorate more readily than others (eg organics, including early plastics). For some materials there are apparently successful treatments (eg the use of corrosion inhibitors on excavated bronzes), while for others none has yet been developed. Many early treatments have proved to be less effective than was hoped (or have even proved to be damaging) and some treatments in current use are not yet proven.

Putting the conservation proposal into action

Doing nothing

In some cases the decision will be made to do nothing. One situation in which this decision may be taken arises when it is simply not practicable to conserve, other than possibly in the short term. This was the case with many of the medieval and Roman waterfront structures revealed along the bank of the river Thames in London. Building work would have destroyed the structures but constraints of space (both for treatment and, later, for storage and display) and of other resources including money and staff, meant that it was impossible to conserve in the long term. So emphasis was put on conserving *in situ* (by keeping wet while exposed) for long enough to gain as much information as possible from the structures. In other cases wooden structures may not be under threat from building work but in such poor condition that it would not be viable to do more than record and monitor (eg many timbers appearing in the intertidal zone along the edge of the Thames) (Milne *et al* 1997) (Fig 26).

Figure 26 Doing nothing: Bronze Age posts, part of the structure of a timber bridge over the Thames, London. These have been surveyed, recorded and monitored but left *in situ* (Courtesy of Gustav Milne).

More frequently, the decision to do nothing may be taken because, other than maintaining adequate preventive care, nothing more is needed, or because other options might be seen to carry little benefit and too much risk. A particular situation in which no conservation treatment may be undertaken is when the conceptual significance of the material is such that it is considered inappropriate to do more than provide preventive care. The growing understanding of conceptual significance has changed the conservation approach to human remains, and to secret or sacred objects. An interesting example arose at the Canadian Conservation Institute over the proposed conservation of a rare and very deteriorated Mi'kmaq prayer book. Initially a representative of the Mi'kmaq who was helping the conservator felt that it would be inappropriate to apply any remedial treatments because of the spiritual nature of the book. Eventually the conservator flew to Newfoundland to explain possible conservation processes and consult the Mi'kmaq community, and it was decided to treat the book, which has now been returned to the community (Hanington 2000).

It is not only ethnographic material which may be left unconserved. Conceptual artists may prefer their works to be left alone, indeed many artists of different periods may have anticipated deterioration as part of the 'life' of their works 'with certain works, preservation is interference – interference with a process which has been dictated (or tacitly accepted) by the artist' (Fry R 1997, 56).

Change of use

There may be obvious reasons for changing the way in which an object is used, for example to minimize wear and tear caused by handling or over-visiting. Thus one possible objective of changing use may be to safeguard the material integrity of an object or structure. Another objective may be to reinstate conceptual significance as when an object is cared for in traditional ways in a museum or is lent for use in ceremony (Clavir *et al* 1986, Welsh *et al* 1992, Clavir 1994, Simpson 1996, Paterson & Greenfield 1998, Peers 1999). Change of use also occurs when artefacts are returned to an indigenous community or when human remains are repatriated so that they can be reburied (Hubert 1989b, Greenfield 1995, Coote 1998, Leggett 1999). Some changes of use are contentious. When 'Seahenge' (a Bronze Age timber circle in the inter-tidal zone at Holme-next-the-sea, in Norfolk, UK) was excavated and removed in order to preserve it, it ceased to be a feature of the local landscape which appeared and disappeared with the tides and became a subject for scientific study. Its removal for analysis and conservation, and possibly for eventual display, transformed it into a 'museum' object and for some people, at least, this change of use robbed it of its conceptual significance (Miles 2000) (see also Chapter 10).

An example of an object with dual uses, as both museum object and ceremonial object, is the Lord Mayor's coach which now spends most of its life on display in the Museum of London, but leaves the museum to be used in procession on ceremonial occasions. Its use in ceremony requires that it is maintained in working order so that it can carry passengers, and be pulled by horses, while in the museum preventive care ensures that the whole remains in good condition (see Figs 10 and 11).

Preventive conservation

Preventive care

Preventive care is now seen to be vital to the management of change and the long-term conservation of individual objects and collections. In any well-managed organization all objects will be given basic standards of preventive care which means ensuring that they are handled safely, that stored material is kept in well-regulated and clean areas with appropriate environmental control, and that objects on display are as far as possible protected from exposure to touching, excessive light-levels, and so on. Any object which is given some form of preventive or remedial treatment will also be given long-term preventive care. Thus preventive care should be applied to *all* objects which are held in a museum (and to historic monuments).

This form of care involves appropriate control of all the external factors which affect type and rate of change: light, temperature, relative humidity and pollutants (MGC 1992 onwards, Dollery

1994, Erhardt & Mecklenburg 1994, *Museum Practice* 1996 onwards). It also involves minimizing stress which might otherwise be caused by inadequate support or vibration, or by poor handling (Bradley 1990, Rose & de Torres 1992, Robinson & Pardoe 2000). Application of preventive care, without remedial treatment, involves accepting the current state of an object, and presupposes that the conservator is confident that changes to packaging, handling or environment alone will ensure reasonable future stability. The benefits of preventive and remedial treatments will be maintained through follow-up preventive care (La Rocca & Nardi 1994).

Local environment

Preventive care in a museum or historic house should take place within a framework of environmental management for the building as a whole, but preventive care of individual objects (or groups of objects) may involve 'local' management, eg through controlling the environment in specific parts of a building, or in individual storage boxes (Fig 27). Achieving better conditions involves assessing the state of the objects or collections, identifying causes of deterioration, then deciding on an appropriate environment and preventive care regime. Adjustment may mean modifying the environment or changing policies on handling or other aspects of use. It may mean moving an object from one environment to another, eg from outdoor to indoor, or from open display to display in a show-case. For example, some medieval stone sculptures at Lincoln Cathedral were removed from the exterior of the building and replaced by replicas, in order to protect what remains of the original carvings (Price 2000a). Excavated wooden structures such as the Rose Theatre (the remains of Shakespeare's theatre in London) or waterlogged boats have been reburied in order to recreate an environment which will prevent further deterioration (Wainwright 1989, Fry M F 1996).

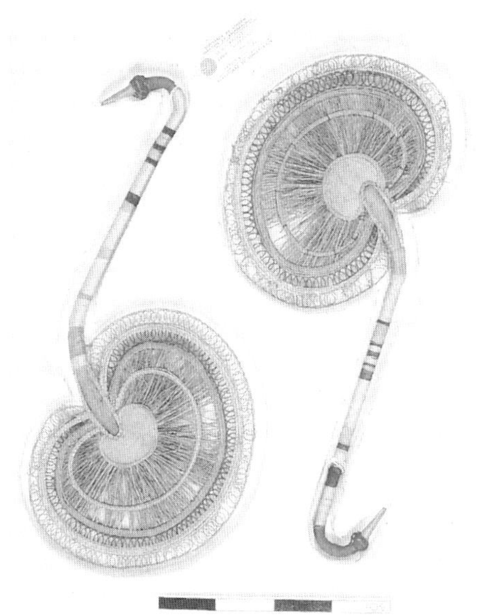

Figure 27 Preventive care: supportive packaging for two fragile ethnographic plant fibre fans (Courtesy of UCL Institute of Archaeology).

Preventive treatment

Preventive treatment was defined earlier as treatment intended to prevent future change while not necessarily remedying past damage, so less intrusive than applying the full range of remedial treatment. The clearest examples of preventive treatment involve removal of materials which are causing on-going deterioration (or may cause future deterioration), including processes such as elimination of insect pests through heating or freezing, or elimination of soluble salts which, if left, may cause disintegration of the object. Some 'cleaning' processes can be seen as preventive in that removal of dust and dirt may not only 'improve' the appearance of an object but at the same time remove sources of future deterioration, such as mould spores (Fig 28). Preventive treatments might also include removal of (recent) metal dowels if they are corroding and disrupting the structure of the object (whereas original metal dowels, as found in classical statuary, would normally be retained because of their significance, and some other means would be found for controlling corrosion).

Figure 28 Preventive treatment: removing superficial dust from an ethnographic plant fibre hat (Courtesy of UCL Institute of Archaeology).

A minimal repair may play an important part in preventing further deterioration, such as securing loose tesserae at the edge of a mosaic before greater loss occurs, or securing paint flakes before they become detached. Applying a support to a weak and fragile object such as a painting or textile is part remedial and part preventive (Durham 1995, Morgan & Cruikshank 1995). Another form of preventive treatment involves removing stresses. For example, a textile or mat with very hard creases or folds resulting from poor storage may be humidified to allow the fibres to relax and the creases to be removed, so preventing breakage of fibres along the line of stress. Regular maintenance of working machinery could be considered to be preventive treatment. 'It is recognised that preventive maintenance of equipment, such as cleaning and lubrication, can prolong utility and avoid costly repairs' (Williams 1997, 198).

There is one situation where changing the *structure* of an object could be described both as a preventive measure and a way of enhancing use. This is the case of firing unfired clay cuneiform writing tablets which are badly affected by damaging salts. The salts cannot be soaked out of the tablets in the normal way because the water would soften the clay and cause the tablets to disintegrate. Firing permanently changes the clay structure but makes the tablets resistant to water and thus allows removal of the salts. This is an example of a treatment which would not normally be considered but which makes it possible to handle the tablets and read the texts. It is worth noting that it was Rathgen who first devised this treatment more than a century ago, and that it is still in use today (Organ 1961).

Excavation and preventive treatment

During excavation objects are exposed to an abrupt change in environment (eg an object in damp soil is exposed to sunlight and drying winds; or an object in an oxygen-free deposit is exposed to air) and, at the same time, the soil and adjacent objects, which have provided physical support in the ground, are removed, leaving weak objects very vulnerable. Thus the first conservation process that an archaeological object may undergo is excavation using conservation procedures intended to prevent unnecessary collapse or disintegration.

Figure 29 Preventive care during excavation: a post-medieval fish-trap revealed during excavation and lifted by supporting it on polyurethane foam (Courtesy of Museum of London).

Many objects are, of course, in reasonably robust state and once partially revealed can be exposed further, rapidly assessed and removed safely from the surrounding soil. However, exposure of an object may reveal that it is already very fragile, or likely to deteriorate rapidly (eg waterlogged wood). Before anything more can be done it may need emergency treatment. This is a form of 'first aid' conservation and may involve the use of temporary adhesives (to hold a crumbling ceramic together), or special protective or storage conditions (eg keeping waterlogged organic materials properly wet) (Watkinson & Neal 1998).

Some objects are so complex, fragmentary or fragile that they cannot be removed without providing additional protection during the process of excavation. In this case they may be given exterior support (eg bandaging or polyurethane foam) in order to hold the whole together while it is released from the ground (Fig 29). Lifting in a block maintains the three-dimensional relationship between different objects in an assemblage and enables conservators to excavate the block and record relationships in the conservation laboratory. An alternative approach with structures such as kilns or mosaics is to partially dismantle and remove them in well-documented and manageable pieces (Sease 1987, Payton 1992, Watkinson & Neal 1998). Some of these processes are feats of engineering, such as the release of the mass of lime-plaster figures from the site of 'Ain Ghazal in Jordan (Tubb & Grissom 1995, and see Fig 38) or the dramatic lifting of the *Mary Rose* from the sea bed in the UK (Rule 1982).

The general principle is that the shock of uncovering and moving the object should be minimized as far as possible by using appropriate materials and techniques, and that conservation treatments should follow. The exposure of the object may set up a succession of deterioration processes leading to loss of information, and possibly loss of the object, unless assessment and conservation go ahead as soon as possible. An important issue is that, as excavation is potentially damaging, it is unethical to remove objects from an unthreatened site unless provision for conservation (and use) is assured (though it may, of course, be necessary to 'save' an object or structure during a rescue or salvage excavation).

Where it is inevitable that building development will damage or destroy important remains, means are sought to minimize the impact, eg through modification of the building design to leave most of the remains accessible, or through excavation and removal for study and display. The responsibility for mitigation of the impact of development on the site and its contents is increasingly placed on the developer. In the UK this is now built into the development planning process (UK Department of the Environment Planning Policy Guidance 16: Archaeology and Planning, 'PPG16').

Is preventive conservation better than remedial conservation?

Codes of practice (eg AIC 1996) stress the importance of preventive conservation.

> *The conservation professional should recognize the critical importance of preventive conservation as the most effective means of promoting the long-term preservation of cultural property. (AIC 1996, 29)*

> *Preventive conservation is more than a physical activity. It is consistent with philosophies and practices that are basic to ethical standards endorsed by the museum community. (Williams 1997, 199)*

The implication in both these quotes is that preventive conservation is better than any other approach to conservation. While not disagreeing with the view that preventive conservation is crucial to the preservation of museum collections, it is important to examine whether preventive conservation is always 'better'.

If preventive conservation is taken to mean procedures which do not involve direct interaction with objects (what has been called earlier 'preventive care'), there are some clear advantages because the procedures used should not change the material significance of the object, thus it can be said to satisfy the criterion of minimum intervention. In terms of efficiency of managing change, preventive care is efficient *only* if objects have been properly assessed and given an appropriate environment, and it remains efficient only if the environment is monitored, and modified if necessary. Assessment should have revealed whether an object requires some direct treatment before it can be stabilized in an appropriate environment. Only if this form of preventive conservation is efficient will it safeguard the significance of the object.

Preventive care can be said to be better because it is assumed to be cost-effective, in that it should obviate the need to remedy the effects of deterioration on individual objects (Staniforth 1990, Keene 1996). Using this approach to conservation it is possible to minimize change in whole collections rather than deal with objects individually. It should remain cost-effective provided it is effectively monitored, but if it is not monitored and the environment fails (allowing a major problem to arise such as an insect infestation, or an outbreak of corrosion on metal artefacts), large-scale and expensive remedial measures may be needed. Monitoring the environmental conditions, eg light levels and relative humidity levels, does not require specialist conservation expertise but monitoring the effects of the environment, and proposing modifications, does.

It can be said that limiting conservation to both preventive care and preventive treatments is 'better' than undertaking remedial treatments, since even applying preventive treatments is less intrusive and less expensive than also applying remedial measures (so, for example, treatment of excavated ceramic can be limited to removing potentially harmful salts rather than also rejoining the sherds in order to remedy the damage caused by breakage or crushing). Preventive care does not make objects readily accessible as it is at its most effective in well-monitored storage areas where it is possible to exclude light and dust, and buffer against vibration and humidity changes with suitable packaging, but in that environment material is neither visible nor easily accessible (see Fig 20).

Prevention alone may not render an object useful either to the public or to specialists, indeed it may be thought to 'sterilize' it. If the intention of conservation is to enable objects to be interpreted and used (thus aiding access to them), then it is almost certain that some remedial treatments will be needed. The point of elucidating the significance of an object is to make it available for study, or teaching, or for demonstration or display, but preventive conservation cannot necessarily do this since it normally results in maintenance of current use-value. This is in contrast to remedial conservation which should result in enhancement of use-value. In many cases remedial and restoration skills are vital tools for understanding objects and aiding interpretation, as in the example of excavating a mass of objects lifted in a soil block to reveal the fragile remains of a trinket box and its contents (see Figs 30–33). Thus although preventive measures are undoubtedly important from both the ethical and the practical points of view, the possibilities of gaining information and of using objects would be limited without remedial and restoration measures.

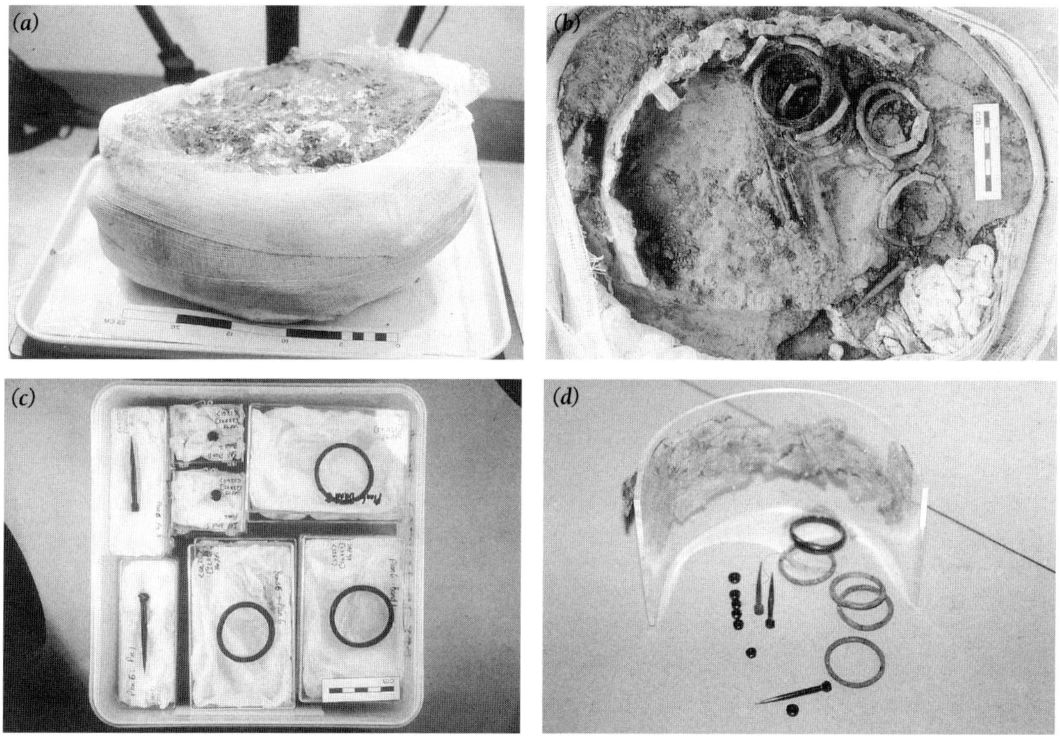

Figure 30–33 Remedial conservation resulting in gain of information. A Roman trinket box after lifting in a soil block supported with bandaging (a). The contents being excavated and recorded (b). Pins, beads and bangles made of jet, shale, cannel coal and horn after investigation and conservation (c). The remains of the fragile copper alloy box and its contents on display (d). (Courtesy of Museum of London).

Assessment of excavated archaeological objects (particularly recently excavated ones) in terms of condition and potential for providing information is often impossible to do without investigative work on the object itself. This may be limited to discovering what kind of object is beneath an encrustation, and the object may be stored thereafter in a controlled environment, but investigative skills will be required to reveal significance. When dealing with existing collections, removal of old repairs or restorations which may be damaging the material or object, or distorting its conceptual significance, will also require remedial conservation.

Preventive care has the advantages that it should minimize material change and can be used as a tool for maintaining condition in whole groups of objects. In these senses it presents significant advantages over labour-intensive remedial treatments applied to individual objects. However, there is a distinct danger that many museum managers will assume that environmental management is all that is needed to keep collections in good condition, and that monitoring by conservators is not needed. In fact many problems cannot simply be solved by modifying the environment, and the skills of the experienced practical conservator are still essential, especially when the condition of collections may be taken to be an indicator of performance in preservation of collections (DCMS 2000). Experienced conservators are needed to modify and monitor the environment, and to monitor the condition of the displayed and stored objects, since if control mechanisms fail and the failure goes undetected, lasting damage may be done to collections.

In fact management of the environment, far from being the passive business implied by the term 'passive conservation', is a dynamic activity since the systems of control require to be checked regularly to ensure appropriate conditions are achieved and maintained (Sandwith & Stainton 1991, MGC 1992 onwards, *Museum Practice* 1996 onwards, Pinniger & Winsor 1998). However, the increasing demands for access, new displays and greater communication with the public mean that it is difficult for conservators also to keep up an effective programme of monitoring and care of material in permanent storage.

Remedial conservation

Some problems of poor condition or illegibility can be alleviated by applying remedial treatment. Broadly, remedial treatments can be divided into those that involve removing materials from the object (because they may be obscuring shape or surface detail), those that involve adding materials such as adhesives (to strengthen a weakened structure, or to reconstruct a broken object, or to replace [compensate for] losses) and others which may involve either modifying the shape, eg where an object has been crushed, or changing the chemistry to produce a more stable material. In each case the object will be changed, so there should be a valid reason for undertaking the treatment; remedial conservation should not be undertaken simply because it is possible. It is also important to remember that any treatment which makes a visible change to an object is interpretative and affected by the views of the conservator:

> *despite the trend of minimalism, interventions of some sort are often inevitable and usually, by implication, involve a statement, realised on the object through the actual treatment. (van de Wetering 1996, 197)*

Removing materials

Many objects and structures have accumulations of materials on part or all of the surface, such as dust and oily deposits. These accretions or changes alter the colour and texture of the surface and may obscure surface features and details of decoration or design. Some of these changes are seen as the inevitable and desirable indications of age and use (patina). However, they may make it difficult to interpret the object (in the case of highly encrusted excavated objects) or to assess and understand the condition of an object or structure, and, worse, some accretions may consist of materials which are also involved in active deterioration processes.

Some of the accretions will be foreign to the object (soil on the surface of excavated objects, dust on the surface of museum objects) but some will have been formed as a result of the reaction between the object and the environment (corrosion on metal, changes to the surface of glass). Still others will be a mixture of intrinsic and extrinsic accretions (corrosion on metals incorporating soil and plant material) and many accretions will be indicative of the history or use of an object (blood-stains on a military uniform, carbon on a cooking pot, textiles trapped in corrosion products on a brooch). Thus not only should some accretions not be removed, but removal should be approached with caution because it makes a permanent change and can affect all aspects of significance, and use. Just like excavation, removal of accretions can reveal one form of evidence while destroying another, and it can change the overall appearance of an object very considerably. It is difficult to ensure that only damaging materials are removed; for example washing textiles may remove adhering dirt, mould spores and acid deterioration products, but it may also remove dyes and stains or other deposits which provide evidence of manufacture and use (Brooks *et al* 1996, Eastop & Brooks 1996). Because of the evidence contained in accretions some conservators prefer to undertake partial removal only, for example to reveal part of the underlying surface on a metal object, but this may not always be the best way to ensure long-term stability. The skill of the conservator lies in detailed examination of the object and accretions, careful choice and execution of removal technique and knowing where to stop in order to safeguard the integrity of the object.

'Cleaning'

Removal of materials from an object is often described as cleaning. This is an unfortunate term since it makes the many different actions which can be referred to as 'cleaning' sound like one simple and innocuous procedure. It is also unfortunate because it has connotations of purity and of achieving an object unsullied by 'dirt', when in reality conservators and other stakeholders aim to retain informative accretions. Use of this term may also encourage non-specialists to see it as a procedure anyone can undertake. The procedures which involve removal of obscuring or damaging materials are some of the most difficult to control and the results can be highly controversial. In fact 'cleaning' covers a range of activities from the scientific investigation of archaeological objects to the aesthetic exercise of elucidating a painting. Because the state of the object beneath the stains, concretions, or earlier restoration is difficult to establish, even with sophisticated examination techniques, and because objects react in different ways to conservation treatments, removal of materials is always a 'new' experience and may produce unexpected results. Conservators have to make choices during these processes, thus the activity is subjective and relies on their interpretation of the object.

Removal of materials from museum objects

Removal of materials from museum artefacts is not necessarily undertaken in order to make them look 'clean' but to clarify significance. It can have a number of different effects so the reasons for the procedure may be varied. Many procedures are undertaken to remove deposits or stains which prevent appreciation of the colour, gloss or texture of the underlying surface. Some may be undertaken to locate the surface, and clarify its topography, or to clarify condition by revealing cracks, breaks or earlier repairs. Some, particularly washing techniques, remove contaminants so may prevent further deterioration, and they may also remove stresses by enabling creases or folds to be eased out.

Since the significance of many museum objects depends (partly) on their being old, 'cleaning' involves achieving a balance between clarification of surface features and retention of the signs of age.

> *An ideal restoration should not at first glance be evident. The work should appear in good order, legible in all its parts, but showing also its age. A balanced and careful cleaning has the fundamental purpose of achieving this. (Schleicher 1997, 152)*

Achieving this balance is difficult, and removal procedures are problematic because they make irreversible changes to the object. Although the intention is to regain something of an earlier appearance so as to clarify the meaning of the object (Figs 34 and 35), procedures can remove not just the signs of age but evidence which may aid present or future interpretation.

Figures 34 and 35 Remedial conservation resulting in gain of information: early nineteenth-century lead coffin plate before treatment and after removal of corrosion to reveal the inscription (Courtesy of Museum of London).

For example, at the root of some of the controversies over the 'cleaning' of paintings is the knowledge that subtle shading or gradation of colour consisting of thin paint layers or transparent coloured glazes can be removed during the process (Hedley 1990). It may also result in the removal of materials which have not been detected, either because they have not been looked for, or because there is no means yet of revealing their presence (Fig 36). This problem is particularly likely to occur when dealing with archaeological material where evidence such as lice on a comb, or blood on stone tools are difficult to detect and could easily disappear while removing other materials (Fell 1996, Wilson *et al* 1996).

Figure 36 Potential for loss of evidence: a nineteenth-century Maori wooden ladle, possibly used for mixing red ochre and oil for body painting. Conservation treatments could result in the loss of important traces of oil and pigment (© Copyright The British Museum).

Although 'cleaning' may result in clarification of colour because overlying veils of dirt or salts have been removed, it is important to distinguish between this situation and colour change which is inherent to the object. Fading of pigments in paintings or natural history specimens, darkening of some woods in furniture, or yellowing of paper cannot normally be improved. Taking removal of stains or accretions too far can result in too bright an object and loss of the 'magic' of age; this can be particularly problematic on a composite object since the result may be that some components appear brighter than others and so affect the whole balance, or accentuate differences in condition.

Inadequate knowledge of the materials and technique of the object, or of the effects of deterioration, can lead to poor choice of removal method. This was the case with the thirteenth-century Norwegian Haug crucifix which was mistakenly assumed to be painted in egg-tempera when in fact the medium was oil. Removal of later overpainting, rather than leaving the original paint unaffected, caused devastating damage (Kollandsrud 1997). This can be compared with the loss of information caused by indiscriminate electrolytic or electro-chemical treatment of archaeological metalwork (commonly practised in the first half of the twentieth century) which stripped off all corrosion and much surface detail. In this case the damage was caused by the failure to appreciate that much of the surface information lay within the corrosion layers. Much recent research into 'cleaning' procedures concerns the development of processes (such as laser cleaning, or the use of gels) which enable the conservator to work cautiously by removing one layer of 'dirt' at a time, and limiting the effects of the treatment on the underlying object (Mills & Smith 1990, Cooper 1998). Finally, it is important for curator and conservator to realize the extent to which fashion dictates appreciation of objects and thus their conservation and restoration. Larson's discussion of the early twentieth-century removal of what was presumably original polychromy on Renaissance terracotta busts (to leave a bare terracotta surface) is particularly instructive (Larson 1997, Larson 1990).

All these disadvantages strengthen the argument for a cautious approach to removal of dirt or accretions, and for partial treatment where this is appropriate. However, provided the conservator has good understanding of the object, and provided the process is skilfully and carefully undertaken, it can have undoubted advantages (Fig 37). Removal of yellowing and dirty varnish reveals colour and detail in paintings, removal of corrosion on excavated metal objects reveals ornament and tool-marks, all of which would otherwise remain invisible and unappreciated. Hunter and Foley discussed the complexities of dealing with the corroded enamel-work of the Anglo-Saxon Lincoln hanging bowl. Their objective was to reveal the

Figure 37 Revealing the image: removal of superficial 'dirt' from a Roman Mummy portrait, clarifying the flesh tones and the modelling of the face, Petrie Museum of Egyptian Archaeology, UCL (UC 14692) (Courtesy of H & R Jaeschke).

original colour of the red enamel, as well as other technological details; they left small areas of corrosion in order to preserve surface evidence for future re-examination but decided that 'when considering a display object of this stature it was important to make accessible to the public and specialists alike the brilliancy and fine detail of the original enamel work' (Hunter & Foley 1984, 18).

Although 'cleaning' has the potential to damage it is one way of understanding objects, just as excavation is one way of understanding a site. If obscuring or damaging materials were never removed many objects would have little apparent significance or use. It is becoming more possible to interpret some objects using non-interventive examination techniques such as X-radiography (just as it is becoming possible to interpret archaeological sites using different types of non-interventive survey), but to detect details and phases of construction, effects of use and modification, as well as to expose the extraordinary range of technological and aesthetic achievements of past peoples, some intervention remains necessary. Thus 'cleaning' is one possible way of making some aspects of the past (and the other) visible.

Removing materials from archaeological objects: conceptual issues

Conservators know that many of their treatments change the appearance of objects. Should consideration of the maker's or user's intent affect what archaeological conservators do? This has not been considered often enough although a recent conference in France opened up discussion (ARAAFU 1996).

> *When nineteenth- and twentieth-century works are conserved, our views are focused not so much on the passage of time but on the ideas and techniques which the artist used to create the work originally. The artist's original intent is rarely mentioned when conserving ancient works. (Heuman 1997, 121)*

In fact, the consideration of artist's intent is always problematic because it is impossible to know the intent of a long-dead artist, and changes in the object have almost always changed the appearance. This point was made by Hedley: 'in paintings which have undergone significant changes, we cannot any longer have the artist's original intention' (Hedley 1990, 9). However, the fact that it is impossible to envisage the intent of the makers and users of archaeological objects does not mean that their possible expectations should not be considered.

This question of maker's intent arises, for example, over the role of colour in ancient artefacts. Some deep red glass becomes green because of the effects of burial on the copper salts which impart colour to the glass. The question then arises whether to retain the green appearance or to reveal the intended red colour, even though this means removal of some of the original glass surface, albeit now altered (Smith 1999). Ancient bronze objects (eg brooches, small figures, tools) may have been intended when in use to be either bright (polished) or dark with a 'tarnish' corrosion layer. The makers would have been familiar with the eventual formation of green corrosion in a neglected or weathered object, but they would probably not have expected the objects to look liver-red or part red and part green (sometimes spotted) which is the effect which can result when conservators remove overlying corrosion to reveal surface features such as engraved decoration. Many archaeological conservators will say that colour is not as important as surface detail: 'we look for an epidermis that is on occasion admirably conserved under a crust of mineral products. Color itself means nothing' (France-Lanord 1996, 247). But should considerations of colour be sacrificed to reveal surface characteristics (which could in many cases be detected with X-radiography)?

Different standards of 'cleaning' clearly produce different visual and conceptual effects. To consideration of maker's intent must be added the conceptual, or aesthetic value for today's audience, of corrosion ('patina') resulting from burial. It would not be surprising if museum visitors assumed that most excavated copper-alloy (bronze) utensils were green when in use, since they are commonly displayed with a green corrosion layer, so is this misleading? For some people, however, the patina not only indicates the history of the object but may embody additional aesthetic significance. Thus conservators should consider the effects of removal or retention of accretions on the possible interpretations of the historic evidence, as well as on the appreciation of the object for its aesthetic qualities and its age-value.

In fact, there are inconsistencies in visual presentation of corroded ancient metals. Where the condition allows, silver is frequently polished to provide a brilliant shine, but bronze, even when it is in a condition which would allow it, is not. Highly polished silver may lose any signs of history and age, as was apparent at the 1989 British Museum exhibition entitled 'The Work of Angels', in which a range of Celtic metalwork from several different museums, each with its own approach to 'cleaning' (or not 'cleaning') its objects, was exhibited together. Some of the highly polished silver looked as if it could have been made yesterday, and this is apparent in the photographs in the catalogue (Young 1989). Polishing silver may be imposing today's values, as little is known about the appearance of silver during its original use, though it has been proposed that, during classical times, not only was silver allowed to acquire a smooth black stable tarnished surface but that red and black figure ware are representations (skeuomorphs) of gold and silver vessels (Vickers 1995). There are no clear answers here for the curator and archaeological conservator but these issues should be considered and discussed.

Adding materials

Another approach to prolonging the use of an object is to add materials to strengthen a weakened structure, or to join dissociated fragments or to replace missing features. The issue here is that, although repair and replacement are intended to enhance conceptual significance, adding materials may make just as permanent a change to the material significance as removing materials. A number of different materials may be added, such as corrosion inhibitors, protective coatings, adhesives or consolidants.

Consolidants

Some materials may be seriously weakened by the effects of loss of a component material, eg loss of cementing material in stones. To maintain or regain structural coherence it may be necessary to introduce a material (a consolidant) which will penetrate into the weakened structure and strengthen it. This procedure is used only if absolutely necessary as it is never possible to remove more than a very small proportion of the consolidant, in other words consolidation makes a permanent change to the structure of the material.

Introducing a consolidant can be seen to be partly preventive in that it is intended to safeguard the object against future disintegration, but it can also be seen to be remedial in that it is countering the damage which has already taken place. A good example is the treatment of waterlogged wood which has lost much of the cellulose which strengthens the cell walls. Introducing a material which bonds to the remaining cellulose and fills the voids in the degraded cell-structure, while at the same time allowing water to evaporate from the wood, results in a artefact which can be handled and displayed (Pearson 1987). The extraordinary 9000-year old lime-plaster statues from 'Ain Ghazal in Jordan provide another example of severely weakened objects which could not be studied or appreciated without remedial treatment. They were lifted as a cemented and friable mass and required consolidation in order to separate and reconstruct them. Without this treatment it would have been impossible to study them in any detail, or to discover how they were made (Tubb & Grissom 1995) (Figs 38 and 39).

Figure 38 Consolidation and reconstruction: the crushed fragmentary 9000 year old 'Ain Ghazal lime plaster statues lying in the lifting crate at the start of the process of consolidation (Courtesy of UCL Institute of Archaeology).

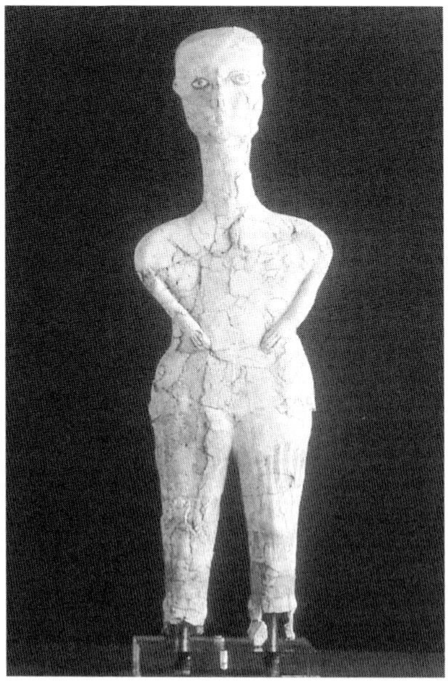

Figure 39 Consolidation and reconstruction: One of the 'Ain Ghazal statues after consolidation and reconstruction for display (Courtesy of UCL Institute of Archaeology).

Repair and reconstruction

Reconstruction contributes to revealing both historical and archaeological evidence and reinstating conceptual significance since the reconstructed whole provides information on size, shape and possibly on original use. Repair or reconstruction may also play an important role in stabilizing an object or structure. A ceramic or glass vessel in fragments may be illegible, and the individual sherds vulnerable to damage if they are not well packaged. Reconstruction can make vessels comprehensible for display (but specialists may prefer to examine sherds in order to see the inside as well as the outside).

The normal method used is to join fragments with an adhesive. Where necessary, for strength, joins are sometimes supported by the insertion of dowels (which involves drilling into the original material, thus involving loss of substrate and possible damage to the object). Repair can also involve filling or bridging gaps in order to strengthen reconstructed areas, for example when parts of an object such as a ceramic vessel are missing. Even though the intention here is to strengthen, filling any gap has the effect of partially completing the shape and so compensating for losses.

In some cases original structural material is so weakened that it is removed and replaced rather than strengthened. This is the approach often used in buildings where individual components such as beams must continue to bear loads (and in such cases an insect-eaten wooden beam might be replaced rather than consolidated). Similarly, in order to maintain a weather-proof outer surface deteriorating exterior stonework may be replaced at intervals as part of a normal maintenance programme on an ancient building.

Modern adhesives

The use of natural adhesives such as bone glue has a long history, but at about the same time that conservation was evolving as a recognizable discipline (in the 1930s), modern synthetic polymers (plastics) were being developed. When these came onto the market in the form of adhesives they appeared to offer significant advantages because they were more versatile and less visible, when applied to objects, than traditional materials. However, the subsequent discovery that many of these adhesives had poor long-term properties and that many are difficult to remove has led to a much more cautious approach to their use (eg Sease 1981, Hanna 1990).

A significant problem for conservators is the fact that modern adhesives are formulated for industrial applications which seldom match conservation requirements. On the other hand it is an advantage that there are many available adhesives and conservators can test and select from a wide range (although there are limitations to the reliability of short-term testing). Conservators look for an adhesive which is weaker than the substrate (the object) so that the adhesive fails rather than causing damage to the object. They also look for compatibility with other factors, such as the flexibility of an object (eg plant fibre in basketry) or the climate in which the adhesive is used (some adhesives fail in high temperatures). The acknowledged instability of many modern adhesives remains a disadvantage and research into adhesives for conservation is now going in two directions. One is to re-evaluate natural adhesives such as starch pastes since their long-term behaviour is at least well-known (Cruickshank & Tinker 1995). The other is to continue the search for modern adhesives with better long-term stability.

The issue of reversibility is particularly relevant to added materials such as adhesives. Although it is not possible to reverse the process of adding an adhesive, most conservators wish to apply only those materials which they expect to be at least partially removable at a future date, so that permanent material changes can be minimized. This was broadly the intention of the principle of reversibility when it was first proposed. However, adhesives often prove difficult to remove because they have become firmly bonded to the object, or have changed with age and can no longer be dissolved, and even when it is possible to remove enough of an adhesive to separate old joins, traces will almost certainly remain. Conservators may choose to use two different adhesives at different positions on a join not only to improve bonding, but to aid future separation.

Reversibility is also sometimes mentioned in relation to consolidants, for example, polyethylene glycol, the polymer used to consolidate and strengthen waterlogged wood, appears to remain water-soluble in laboratory conditions, and thus theoretically it could be removed from treated wood. However, attempting to remove a consolidant from any weakened object would put the object at great risk, thus treatments of this kind are considered to be permanent. In some cases it is possible to keep the added consolidant to a minimum in order to satisfy the principle of minimum intervention. This was the approach used by Daintith (1995) when consolidating powdering pigments on the surface of ethnographic pottery from Papua New Guinea, because it was known that the pottery would be handled only rarely.

Adding materials: compensating for losses

Loss of parts of objects affects conceptual significance, particularly when the loss is great enough to distort the appearance or image. Losses can also be important evidence of history as, for example, the jagged holes in the stonework of buildings such as the Victoria & Albert Museum in London, which are left as a record of war damage (Ashley-Smith 1999a). The intention of completing missing areas and reconstructing missing features (sometimes known as loss-compensation) is to regain an approximation of the original appearance (or function, in the case of replacing moving parts in a working object). This completion may have no effect on the long-term conservation of the object in that it is not necessarily intended to manage change.

Issues arise over the choice of materials used to complete missing features. Some conservators argue that as far as possible the same materials should be used (stone with stone, leather with leather) because only in this way can it be ensured that both original and added materials react in the same way to environmental changes (eg organic materials expanding and contracting at the same rate) thus safeguarding material integrity. They may also argue that this is the only way to achieve a sympathetic appearance. Other conservators argue that missing parts should be completed in different materials because only in this way will it be possible to distinguish the added from the original, and only in this way will it be possible to avoid compromising material significance. Most archaeological conservators use different materials to complete missing areas but these materials are chosen to be compatible with the behaviour of the original. Gap-filling materials such as micro-balloons mixed with a polymer, or paper-pulp, have been designed to put minimal stress on a fragile original (Barclay & Mathias 1989, Podany *et al* 1995). Japanese papers are frequently used, for their strength and compatibility, to repair and support other organic materials such as bark-cloth, basket-work and even some textiles (Wills 1995).

In situations where the object is intended to continue its original function (eg a working object such as a clock with moving parts, or a building which must be structurally safe, or a boat which continues to be used in the water) the same or closely similar materials are normally used in replacing lost or damaged features. In this situation the question arises as to whether original or modern working methods should be used since each will provide a different surface finish and appearance. For example, should a replacement wooden beam be hand-sawn or machine sawn?

Completion of missing areas or features and colouring and texturing the added materials to 'match' the original, is intended to regain conceptual significance and is one of the activities which contributes to restoration. It will not return an object to its original state because use, deterioration and earlier conservation will have changed it but it is possible to regain some of the superficial visual characteristics.

Changing composition

Some conservation treatments make a change to the composition and structure of materials present in artefacts, either to produce a more stable structure, less susceptible to change, or to regain something of the original appearance. Examples include retanning disintegrating vegetable tanned leather with a modern tanning agent (Calnan 1992, Wills *et al* 1992), or treating deteriorating building stone or wall paintings to convert soluble calcium sulphate into insoluble barium sulphate (Giusti 1999). In each case the change in chemistry is intended to induce stability (the latter method is analogous to treating human teeth with fluoride). Chemical treatment to convert blackened white lead pigments to a different but more stable white form, or refiring blackened lead-glazed ceramics, are examples of regaining legibility (Tennent *et al* 1996, Daniels 1999).

Understandably, conservators are reluctant to use such treatments because of the loss of material evidence. However, the objective is to regain conceptual significance, thus to retain the use of the object. Some objects which are deteriorating very rapidly may simply not survive without this kind of intervention, while others may lose much of their meaning unless they are treated. Detailed documentation is intended to ensure that the changes made to the object are recorded and material significance is not misinterpreted in the future. Even though apparently effective, the long-term effects of some of these treatments (eg retanning) have still to be fully evaluated.

Changing shape

Many objects have been crushed or distorted during burial, or use, or through poor storage. In some cases (eg a crushed ceramic) it is possible to reconstruct the three-dimensional shape from the fragments, but a misshapen metal object or a squashed basket may need to be reshaped if its original form is to be understood. There are three reasons for caution: firstly, some objects may have been crushed as part of original use (eg ritual breakage of weapons, or deliberate folds in a garment); secondly, the reshaping procedures may affect internal structure; thirdly, reshaping is interpretative because the aim of the treatment is to induce a temporary malleable or flexible state in the object which allows the conservator to manipulate it into a semblance of the supposed original shape. To these reservations may be added the fact that even the later crushing caused by burial may be considered to be part of the history of the object.

Most controversial is reshaping metals because in many cases some heating is needed which changes metallographic structure. Less contentious is reshaping organic materials (paper, textiles, baskets), but exposing them to moisture to render them flexible almost certainly induces some swelling in the cell structure which may leave permanent change.

Figure 40 Changing the shape: balancing loss and gain of information. A Roman lead curse tablet, before being opened to reveal the names of the accursed (Courtesy of Museum of London).

Figure 41 Curse tablet after removal of corrosion and cautious use of warm water had facilitated unfolding. The evidence of the original folds was lost but otherwise hidden information was gained (Courtesy of Museum of London).

Provided the crushing or distortion is not thought to be a feature of original use and does not contribute important information on the history of the object, reshaping can do much to regain conceptual significance. In some cases the results are extraordinary and transform the understanding of the object, as in the reshaping of Roman crocodile skin armour (Sully 1992), or the restoration of an Egyptian dress (Landi & Hall 1979) (see Fig 4). Here, too, documentation provides a record of what has been done so that any changes detected in the future should be interpreted knowing that a reshaping process has been used (Figs 40 and 41).

Restoration

Any conservation process which contributes to enhancing the visual or functional understanding of an object may be described as restoration; thus 'cleaning', reshaping, filling losses may each have a role in restoration. The objective of restoration is to aid the interpretation of an object, so paintings are restored in order to complete the image and industrial machinery may be restored to demonstrate its function (Keene 1994). Unlike much nineteenth-century and earlier craft restoration, today's restoration follows investigation and conservation of the object. The validity of a restoration depends on the conservator's investigation, and the approach to restoration is influenced by the conservation requirements of the object. The continuum of conservation means that some treatments provide both conservation and restoration; thus supporting a damaged textile on a backing may be intended primarily to strengthen the original, but if the backing is of similar colour and texture it helps to minimize visible gaps and holes and so restore something of the original appearance.

The processes which are most frequently associated with restoration are removal of accretions and completion of losses, though treatments such as coating degraded glass with a polymer to restore transparency (Smith 1999) could also be included. Accretions are removed to reveal surface colour and detail, and missing parts are modelled, coloured and finished to give the appearance of a whole and undamaged object (Figs 42 and 43). Of these processes, compensation for losses is more controllable than cleaning, and in many cases can be largely undone at some later stage. Restoration does not necessarily contribute to long-term preservation in the material sense, but it may mean that an object is treated with greater physical respect because it appears undamaged or is more immediately understandable; in all cases restoration *is* intended to preserve conceptual significance.

Figures 42 and 43 Restoration: gain in legibility. Part of a large assemblage of late seventeenth century blue and white tiles which were dismantled in order to reassemble correctly. Missing tiles were cast and painted, based on evidence of similar tiles in the assemblage. (Courtesy of Museum of London).

Restoration is controversial because it depends on the conservator or restorer's *interpretation* of the object. Even though restoration should never be based on conjecture, different conservators will interpret differently the available evidence on which restoration might be based. Another reason for controversy is that skilfully matched restoration may be considered to be close to faking, indeed, if the continuum of conservation is taken beyond restoration it could be said to reach faking (which employs many of the same techniques and skills as restoration).

The dilemma of restoration

In a frequently-used quote the nineteenth-century French architect, Viollet-le-Duc, said of restoration:

> *Both the word and the thing are modern. To restore an edifice means neither to maintain it nor to rebuild it: it means to reestablish it in a finished state, which may in fact never have actually existed at any given time. (Viollet-le-Duc 1996, 314)*

In fact here he was pointing out the essential problems which arise with all restoration, and perhaps particularly with the restoration of buildings, that the original state has inevitably been changed through age and alterations to the fabric. Thus it follows that a completed restoration represents not a former state but *another* state. If the dirty or damaged state is considered not to demonstrate the significance of an object then the curator and conservator choose to 'improve' that state in order to make the object more understandable. Conservation cannot reverse changes, it can only move the object to a different condition or appearance. Thus curators, conservators and restorers must realize that:

> *simple removal of patina on a bronze or a painting will not recover the object's original appearance, but only uncover the present state of the original material.* (Philippot 1976, 374)

Furthermore, objects continue to change and conservators should manage this change rather than 'freeze' the object into a state of suspended animation.

> *Should our function as conservators, then, be to impose our own perspective concerning durability onto the creative process of the artist? The odd result would be the ideals of the conservator taking precedence over those of the artist, as an attempt is made to permanently lock a work of art into a single moment in time.* (Albano 1996, 183)

Sometimes treatment involves a conscious decision to choose a particular stage in the life of the building or object as a basis for restoration, for example in the treatment of polychrome sculpture (Larson 1992). With complex objects this can be the subject of much debate. It is agreed, however, that restorations should respect the appearance of age so it is inappropriate as well as impossible to attempt to restore to a 'pristine' state.

Restoring to an earlier stage

It may be decided that one particular stage in the life of an object is more significant than the present state, but stripping away all later additions or changes can not only destroy age-value, but remove character without clarifying meaning. A good example is the proposal, in the 1970s, to strip out the eighteenth-century houses built into the ruins of the medieval abbey at Bury St Edmunds, Suffolk, UK. If this had been done it would have robbed the ruins both of later history and considerable charm, and would have left only a rubble shell to represent the medieval work (see Baker 1999) (Fig 44).

Figure 44 Removal of later accretions: loss or gain of meaning? Eighteenth-century houses inserted into the ruins of the abbey at Bury St Edmunds, Suffolk, UK. Would removal of the houses (as proposed in the 1970s) have enhanced the meaning of the ruined abbey? (Courtesy of N D Balaam).

In some cases, the decision to return an object to an approximation of an earlier state is made in order to provide visual information not otherwise readily available. This is sometimes done with costume which, altered to suit a later fashion, retains the stitch- and cut-lines of the original shape (Doré 1978, Brooks *et al* 1994). Industrial machinery and wheeled vehicles which are to be used in demonstration are often returned to a functioning state (Mann 1994b), however a less intrusive conservation approach is now being developed (Newey & Meehan 1999).

Different approaches to, and standards of restoration

An important issue is whether it is appropriate to restore so that additions are not detectable to the viewer. Effects of damage and loss can be masked more successfully on some types of object than others. While losses in paintings can be filled and coloured with paint, and losses in furniture can be repaired and concealed with careful choice of replacement veneer, even the most skilful repair of glass or fine textiles is detectable on close viewing. The approach to restoration depends, obviously, on the significance of the object. A statue from an archaeological context will normally be left uncompleted (eg with missing limbs); however, Molina and Pincevin (1994) discussed some examples of religious statuary where the views of parishioners and priest (the users) influenced a more extensive restoration than might otherwise have been undertaken, including the replacement of a hand and a missing cross.

Where it is possible to match shape, colour and texture (as in paintings) is it appropriate to do so? An exact match makes it impossible for the viewer to detect which is original and which restoration. However, where an object has been completed and the new material has not been toned to match the original (eg a ceramic with white plaster of Paris fills), the distraction of the obtrusive new fill may make it difficult to appreciate the original. Archaeological conservators attempt to steer a path between these two extremes, in accordance with Philippot's statement that the aim of painting a restored area is to reintegrate the visual whole (Philippot 1976). Adopting this approach means that restored areas are detectable on close viewing but are coloured and textured so as not to detract from appreciation of the whole. An 'archaeological' approach to restoration of paintings (frequently applied to wall-paintings) is to use an impressionistic technique which involves building up colour and shading through the application of fine streaks of separate colours (*tratteggio*). Viewed from a distance this should complete the image, viewed close-to the restored areas should be apparent (Philippot 1976, Mora *et al* 1984).

The case of the Lansdowne *Herakles* provides a particularly interesting example of the contrast between an eighteenth-century artist's restoration and a 1970s 'evidential' re-restoration. Eventually the starkly honest (in that it made no attempt to conceal filled areas) twentieth-century restoration was removed and the eighteenth-century one reinstated on the grounds that the 'honest' approach detracted from, rather than enhanced, appreciation of the original (Podany 1994).

Detachable restorations

The principle of minimum intervention has affected the extent of restorations, and the techniques used, and has now given rise to the concept of detachable restorations. Although there are many examples of attached material being removed at a later stage in order to attempt different and 'better' restorations (eg Williams 1992, Podany 1994), the concept of easily detachable restored features is relatively new. The objective is to position the restored feature in such a way that it can be removed with minimal application of solvents or other direct intervention on the object. The sixth-century BC Greek marble kouros at the J Paul Getty Museum was reconstructed and the missing areas filled in such a way that it is 'possible to disassemble the object with ease and without adverse effect on the individual segments' (Podany 1987, 376). Detachable gap-fills in ceramics have been demonstrated by Koob (1987) and non-intrusive detachable upholstery by Balfour *et al* (1999): here one of the aims was to allow access to the frame of the furniture for future examination of technological details.

Another approach which can restore the appearance of a fragmentary object is to locate the fragments in their assumed relationship on some form of support without attaching the fragments one to another (Fig 45). This was the method chosen to reconstruct degraded fragments of a workman's waistcoat (possibly as early as fifteenth century in date) found in a coal mine (Brooks *et al* 1994).

Figure 45 Restoration of shape using an internal support: a pair of medieval boots, one reconstructed by shaping an internal support to the boot and using fine pins through original stitch holes and into the support to hold the leather pieces in position (Courtesy of Museum of London).

Retreatment; re-restoration

Objects frequently need re-conservation and are sometimes considered to need re-restoration. Retreatment may become necessary if an earlier treatment has failed, or has proved unsuitable for present circumstances. The new treatment may 'reinforce' what has been done before (eg reconsolidate a still-friable object), or may counteract an unsuitable method (eg removal of metal dowels which are now corroding). The concept behind the principle of reversibility was that of facilitating retreatment (after removing the earlier applied material) and some have proposed a principle of 'retreatability' (ie designing treatments to allow for further treatment) to take the place of the problematic principle of reversibility (Appelbaum 1987). Conservation processes are not permanent, the object continues to change and added materials change too. In any case 'maintenance' conservation (ie small interventions at regular intervals) is more likely to manage change and achieve minimum intervention over all, than major and intrusive remedial treatments at irregular intervals.

Re-restoration may be undertaken because the earlier restoration has deteriorated (eg painted reintegration has changed colour and become obtrusive) or because the object itself is at risk. This was the reason for the re-restoration of the Portland Vase (the famous Roman cameo glass vase) at the British Museum. The vase had been restored more than once and the latest adhesive appeared to have failed. A new restoration would also allow for the incorporation of small fragments which had been left out of earlier restorations (Smith 1992). More complex is the situation where an earlier restoration represents an unfashionable, or possibly inaccurate, interpretation of an object. Re-restoration, in the sense of a new interpretation, will introduce another view of the object which, in its turn, may fall out of favour in the future (as in the case of the Lansdowne *Herakles*, see Podany 1994). Re-restoration may also remove significant evidence of how the object was seen at an earlier stage in its history so is normally undertaken only if further evidence indicates a different interpretation. The *Laocoön* (a renowned Hellenistic marble sculpture) had been restored more than once during the Renaissance, then in 1905 part of a missing arm was rediscovered. Eventually this original arm was replaced (which changed the image considerably) but the influential Renaissance restoration, by Montosorli, was recorded in a plaster cast (Philippot 1996, Pinelli 1996). Much re-restoration falls somewhere between these examples (eg Elston 1998) but remedial retreatment, or re-restoration, should never be undertaken simply because it is possible, only if there is a valid objective. The Anglo-Saxon Sutton Hoo helmet was re-restored in order to achieve a more accurate interpretation (Bruce-Mitford 1978, Williams 1992) (Figs 46 and 47).

Figure 46 Re-restoration and new interpretation: the Anglo-Saxon Sutton Hoo helmet before re-restoration. This first restoration was considered reliable in its time (© Copyright The British Museum).

Figure 47 The Sutton Hoo helmet after re-restoration. The re-restoration was more satisfactory stylistically and in the way it would fit the head and neck (© Copyright The British Museum).

Virtual restoration

With the advent of digital imaging it is becoming possible to test potential approaches to reconstruction, restoration (or re-restoration) by manipulating the image of the object rather than interacting with the object itself. Digital imaging also has the ability to 'restore' lost colour, so faded colour can be intensified on the image in order to regain the impression of a more-or-less unchanged state. This approach, too, is interpretative but it has the advantages of allowing curator and conservator to produce more than one interpretation and to compare them, and it satisfies the principle of minimum intervention because it leaves the object in its present state.

Replicas

Replicas can be used not only to make objects more readily understandable but as an alternative to restoring the original. Thus they, too, can support the principle of minimum intervention. However, a replica has the problem that it is an even clearer statement than a restoration of what may be only one possible interpretation, and some replicas are so powerful that reinterpretation could well be hindered (even though the original may remain unrestored). Nevertheless, replicas can have a useful role. Sometimes an object is so changed that both colour and detail have disappeared. This is the case with the Sutton Hoo helmet which is now a drab brown because, during burial, the iron of the helm was converted to rusty corrosion products which also affected other metals present. The replica, made in collaboration with the Royal Armouries, and now displayed with the original at the British Museum, shows what the helmet may have looked like when first made. The brilliant tinned, gilded and ornamented surface would have made the helmet an impressive sight. The replica had a further role in that it confirmed that the new reconstruction provided a viable shape which would have fitted the head, and provided adequate protection for ears and neck (which the earlier restoration would not have done) (Figs 46, 47 and 48) (Bruce-Mitford 1978, Williams 1992).

Figure 48 Replication: further interpretation. The replica of the Sutton Hoo helmet (© Copyright The British Museum).

Replicas can be used to construct a whole where only part of an object has survived. The lines of ships and boats can be extrapolated from the incomplete remains of vessels, and replicas (or scale models) can be made and tested to see how the original might have handled on the water (Greenhill 1976). Replicas can also have a role when the decision is made to do little or no remedial conservation on an object. This was the case with a doublet which was not prepared for display but a replica was made in order to show the probable original appearance of the garment (Brooks *et al* 1996).

Effective conservation

Management of change is complex and multifaceted. Effective conservation is dependent on thorough assessment of the object which requires communication and collaboration between conservators and colleagues, or other stakeholders, and it must facilitate the use of the object, not prevent it. Conservation itself embraces a range of activities from investigation to restoration, with preventive conservation being viewed, rightly, as increasingly important. However, while preventive conservation makes a crucial contribution to preservation, remedial conservation and restoration are needed to aid interpretation of objects, so may play an important role in facilitating use.

8 Working relations

Introduction

Conservation of the heritage is a responsibility shared between a number of different professions and public stakeholders, and, as with any collective concern, there are tensions between the various groups involved. Just as there are varied views about the significance of the material which makes up the cultural heritage, so there are different views on how it should be preserved, who should do the work, and to what standards.

The purpose of this chapter is to demonstrate how social and professional factors in institutions or in wider society affect the impact and success of conservation of the cultural heritage, and to show which factors might be changed or improved in order to make conservation more effective and more widely understood and appreciated.

Conservation has existed as a recognized occupation for less than a century and has gradually developed into a profession only over the last ten to twenty years of the twentieth century. Fundamental changes have occurred within the working lives of many of today's conservators. Equally, major changes have also occurred in museums and in archaeology which have affected the role of curators and archaeologists. Field archaeologists have had to adapt to new and different financial and organizational pressures, curators have seen their traditional wide-ranging role fragment into several new ones, and conservators have taken on wider roles. Each group has had to adapt to a much less certain world requiring a flexible work-force and new skills. During the last quarter of the twentieth century, each group was concerned with defining or redefining its own identity in reaction to these changes, and this tended to create barriers rather than foster collaboration. Conservators, in particular, gained the reputation for being inflexible over the care and use of objects, perhaps because they were attempting to assert their identity in an environment where other professionals such as curator, archaeologist and architect were already established.

Much of this is now changing as the different professions mature and gain greater confidence in their own roles, and as there is increased respect for each other's skills and support for each other's objectives. However, some of the 'baggage' resulting from the old tensions still remains and can still affect working relations, particularly when new tensions arise. It is worth reviewing the social and professional context of conservation as a means to making further improvements.

Changes in the organization and management of field archaeology and museums

In the nineteenth century curators and archaeologists were expected to be both scholarly and extremely versatile. They worked with, and for, a small and knowledgeable audience, often through learned societies. Curators dealt with all aspects of the collections in their care, from study to repair and display, and archaeologists (who were often also curators) had wide-ranging antiquarian interests. During the early twentieth century the practice of curatorship and archaeology gradually became more formalized, partly through development of method (eg excavation techniques, and techniques for restoration and display in museums) and partly through university courses (eg at the University of London Institute of Archaeology, founded in 1937) and professional training (eg through bodies such as the UK Museums Association which provided training from 1934).

Field archaeologists

Field archaeology saw very significant changes in practice and intensity in the second half of the twentieth century. In the UK, redevelopment of city centres (initially after damage during the Second World War, 1939–45, and more extensively after about 1960), and agricultural changes, which encouraged more extensive and deeper ploughing in the countryside, caused alarming destruction of archaeological sites. A detailed picture of the state of archaeology in the UK in 1973 at the beginning of the 'rescue era' is given in the volume edited by Rahtz (1974) and subsequent developments are outlined by Merriman and Swain (1999) and by Hinton (1999). The pressures of rescue archaeology and the growth in regulation meant that, whereas excavations had been run by a small number of field archaeologists based in museums and universities, or through local societies, and with largely volunteer work-forces, now professional archaeological units were formed throughout the country, mainly based outside museums and staffed by graduates in archaeology. These units dedicated to field archaeology saw their primary role as being the recovery, analysis and publication of the information retrieved through excavation, although some (eg the York Archaeological Trust) were also concerned with conservation and interpretation to the public. At the same time different aspects of archaeological practice became established as distinct specialisms (eg environmental archaeology, geophysical surveying, archaeometry) and each developed its separate expertise. New fields of study developed, for example the growing appreciation of the industrial heritage gave rise to industrial archaeology which, in its turn, had an effect on the growth of museums.

In the UK during the last quarter of the twentieth century, paying for rescue archaeology became increasingly the responsibility of the developer (ie potential destroyer) of the site. Consequently, greater emphasis was placed on the costs of carrying out the work, and excavation and the provision of archaeological advice became a much more competitive business. A corollary of this was that the links between archaeological practitioners and specific geographical areas, which had been a feature of local museum-based archaeology, began to break down as commercially-focused archaeologists sought work both across the country and internationally. At the same time working practices became more strictly regulated in terms of matters such as health and safety, and specification of procedures, funding and outputs (Andrews 1991). Professional associations were formed (such as the Institute of Field Archaeologists in the UK) whose role was to set and regulate standards, devise codes of practice (IFA 1999) and provide support for members, and by the end of the twentieth century the archaeological profession covered a wide range of different specialists, including conservators (Aitchison 1999). Similar developments were happening elsewhere, for example in the USA the pattern was paralleled with growth of salvage archaeology, the development of professional bodies such as the Society of Professional Archaeologists (SOPA), and the advent of archaeological contractors competing for archaeological projects.

Museum curators

The role of the curator also changed considerably during the last quarter of the twentieth century, and the effects of these changes on the morale and responsibilities of the curator have been widely discussed (eg Cannon-Brookes 1988, Pearce 1990, Boylan 1993, Kavanagh 1994, Merriman 1997, Merriman 2000). The focus of museum work changed from study of the collections to meeting the expectations of visitors, and from scholarship to education and enjoyment. High visitor numbers were seen as a mark of success, so the range of activities in museums increased, with the result that the old versatile curatorial role fragmented into several different specialisms, such as exhibition designer, education officer, registrar, conservator, even marketing manager and fund-raiser (Sola 1992, Merriman 1997). By the end of the twentieth century it was possible for Glaser and Zenetou (1996) to identify 30 different professions in museums. Boylan (1992) indicated that between the 1960s and the 1990s the proportion of curators to other museum professionals declined from 35% or 40% to 14%. During this period university training in museology or museum studies expanded to provide knowledge not only of the practicalities of museums, such as management and documentation of collections, but also the theory and history of museums, so moving away from the old object-focused training.

Traditionally the curator was responsible for providing suitable conditions for the storage and display of objects, but the poor condition and management of some national collections became a matter of public concern (eg Bourn 1988, Sheldon 1988). Thus the 'scholar-curator' working in a large museum, who considered s/he had an essential role in conservation, saw this aspect of his/her work taken over by a conservator or collections manager (Cannon-Brookes 1988, Cossons 1992, Murdoch 1994). This left him/her as a scholarly object specialist, and with the apprehension that collections management was not only undermining the status of scholarship, but taking an over-large share of the funding. In other parts of the world the poor condition of collections also had direct effects on the development of conservation and management of collections. For example, in Canada it led to the founding of the Canadian Conservation Institute (Clark & Weaver 1982), in the Netherlands to the Delta Plan (Talley 1999) and in Sub-Saharan Africa to the development of the conservation management training (PREMA) programme through the involvement of the International Centre for Conservation in Rome (ICCROM) (Godonou 1995, Antomarchi & Little 1999).

A problem particular to archaeological curators arose from the separation of field archaeology from museums. Archaeological units were responsible for excavation and for the generation of the finds, samples and documentation which constituted the archaeological archive. The archives from many excavations were ultimately deposited in the relevant museum, but the curator had little control over the rate or quantity of this influx of material, and shortage of space and the costs of storage became serious issues. These archives were intended to provide 'preservation by record' and to form a body of material for future study and public use, but by the late twentieth century it was clear that they were under-exploited and represented a major storage and conservation problem (Merriman & Swain 1999).

Towards the end of the twentieth century a further impetus for change was pressure on museums throughout the world to become even more accessible and to respond to wider and more varied audiences (Pearce 1990, Simpson 1996, Specht & MacLulich 1996, Merriman 2000). This was reflected in a change of definition of a museum used by the UK Museums Association. Since 1984 the definition had been:

A museum is an institution which collects, documents, preserves, exhibits, and interprets material evidence and associated information for the public benefit.

In 1999 it was revised to:

Museums enable people to explore collections for inspiration, learning and enjoyment. They are institutions that collect, safeguard and make accessible artefacts and specimens, which they hold in trust for society.

The fact that the 1999 definition started by setting out the benefits of museums to the wider world was a clear indication that museums had shifted their focus from internal concerns to the importance of how they were received and used by the public (Besterman 1998, *Museums Journal* 1999). For curators the balance shifted from safeguarding collections for the future ('for posterity') towards using them to provide activities to attract and interest the public of the present day. Yates, writing about her career as the curator and director of a local museum in Norfolk, UK, pointed out that in the twenty years since 1976 her job had become almost completely involved with enabling different groups of people to use her museum in some way 'My title has not changed - I am still the Curator - but the whole emphasis of my job ... has shifted radically ... My opportunities to handle objects and to study them are few and far between' (Yates 1993, 41). However, it is possible that the balance may shift back to some extent as it has become apparent that the skills of handling and studying collections have become undervalued.

By the end of the century the range of changes and challenges and the diversity of professions working in museums, each with its own ethical principles, had undermined the concept of a single museum discipline or museum profession (Weil 1994). In the UK there was discrepancy in funding and working conditions between major national museums and regional museums.

Twenty years ago, curators in British regional museums were part of an international research community. That has completely stopped. Numbers have been dramatically reduced, and those that are left simply haven't the time to do the research. (MacGregor, quoted in Lister 2000)

Curators felt pressurized, undervalued and uncertain of their role:

This embattledness, which many curators today feel, is because they are not doing the key jobs any more, and museums are not helping curators to redefine what they do. (Cossons 1992, 140)

All these changes happened relatively quickly, driven in some areas by reduction of public funding (eg for many local museums) and everywhere by growth in demand for access to, and use of, the material heritage. There was an increasing requirement to compete for funding from a variety of different sources and to respond to the interests of the public. Glaser and Zenetou (1996) saw museums as entering a period where they would be at the forefront of social change. Not surprisingly, museum curators responded to these rapid upheavals by concentrating on their own position and role rather than on the changing position of colleagues, including conservators.

Issues affecting conservators

There are several issues arising from the perceived role and status of conservation which cause tensions between colleagues in the various museum and archaeological disciplines and which diminish the effectiveness of conservators. Many of these are now easing or changing, but the memory of the past difficulties can still affect working relationships. The different professions have different perceptions; for example conservators have tended to think they are undervalued, whereas in some ways their status has risen, certainly in comparison with curators who may feel their power and influence has been undermined. The relative status of curators and conservators does, of course, vary from museum to museum, and in some museums facing funding and other pressures the status of conservation has suffered.

Conservation as a service

One difficulty is the fact that because conservation has been seen as a service provided to the curator (acting on behalf of the museum and the collection) there is an implied academic (and possibly also social) inferiority. In practice, the conservator is frequently put in a subsidiary position in terms of decision-making, although s/he is usually best qualified to make decisions which affect the preservation of the objects (Corfield *et al* 1989). Conservation does, in fact, provide a service to the heritage and to the owners and stakeholders of the heritage, as indeed do curators, but the provision of a service need not mean that conservation is little valued. Nor should the provision of a service be taken to indicate a lower academic status considering that well-established professions (eg both medicine and law) provide essential services, and at the same time are respected as academic disciplines in their own right.

Another aspect which may have contributed to the hierarchical problem is the fact that many conservators are women, while, until comparatively recently, many of their colleagues (eg curators, field-archaeologists, scientists) were men. Blake (1999), in discussing the reasons why few women become museum directors, pointed out that domestic responsibilities and differences in attitude and confidence can affect the careers of women in museums just as they can in other spheres (and these issues, of course, affect all female staff).

Lack of recognition for manual skills

Another difficulty is presented by the lack of recognition and reward still accorded to manual skills, placing academic skills 'above' manual ones. This is not a new problem. The assumed superiority of using the mind rather than the hands was disputed, for example, by painters writing about their work as early as the seventeenth century, but the problem still persists and affects the way the technical study of paintings is viewed by art historians (Carlyle 1998) or the conservation of museum objects is viewed by curators. This dichotomy is reinforced by the fact that many or most conservators now in senior positions have been promoted to be managers or administrators and no longer work on objects. But when a great work of art such as Leonardo da Vinci's cartoon *The Virgin and Child with St Anne and St John the Baptist* is badly damaged, it is the acknowledged manual skills of a highly experienced senior conservator which are needed to restore it (Harding & Oddy 1992). Perhaps because of this lack of recognition of skills, and consequent lack of authority, conservators are seen as ready targets, eg in controversies over the care and restoration of artefacts such as the Sistine Chapel ceiling, or the Elgin Marbles. However, everyone would, presumably, agree that it is essential that a surgeon or a dentist has highly developed manual skills as well as the required medical knowledge to perform an effective operation.

Lack of a professional identity and voice

The fact that conservation has developed, and changed, from its origins in the crafts of repair and restoration has meant that it has had to attempt to establish a new identity which embraces the academic and the scientific. This has not been easy. There is still uncertainty (even among conservators) about whether conservation is a routine technical service, or an intellectually rigorous discipline, whether it is an art or a science, whether it can make a direct contribution to the interpretation of an object or site, or to the way in which museum collections can be used by scholars and the public. Compared to natural heritage conservation, it remains little known and understood by the public, although, increasingly, museums are putting conservation itself on display, or at least putting more emphasis on explaining the reasons for conservation processes which affect displays and interpretation (such as control of light levels in exhibition areas).

There is a range of different standards of and approaches to working practice (Corfield *et al* 1989). Conservators working in museums, or private practice, at the beginning of the twenty-first century are still drawn from a number of different training backgrounds including traditional apprenticeships, art-schools, in-house museum training and universities. There has also been a growth of different conservation specialisms (eg conservation of paintings, leather, ceramics) each with its own methods and techniques.

The greatest variation in the standards and approaches is probably seen in the extent to which science is embedded in practice, the emphasis placed on aesthetic and restoration skills and the extent and nature of documentation (ie records of treatments given to individual objects and collections). At one end of the scale there are conservators who work in a more or less traditional way as highly skilled restorers with little scientific input and sometimes with little emphasis on documentation, whereas at the other there are, for example, archaeological conservators who put considerable emphasis on investigation, scientific analysis and recording, followed by minimal remedial work (and preventive conservation, rather than restoration). In between there are scientific conservators and restorers who practice the full range of remedial and restoration skills and produce detailed documentation. Until recently there has been no professional requirement that people practising as conservators should meet particular standards of practice, nor that they should update their education and training (and some, of course, were trained twenty or thirty years ago when both practice and training were rather different).

Many conservators have felt pressurized to become more scientific as a part of the process of gaining recognition, to wear the white coat instead of the brown overall or the artist's smock, to work in a laboratory rather than a workshop or studio. This may not have served the discipline well since it may have alienated those rooted in the humanities by appearing to be over-scientific, and the scientists by not being scientific enough. The problems imposed by the lack of a common language between colleagues have been commented on frequently (eg Jedrzejewska 1983). Both Ward (1986) and Drysdale (1999) have suggested that conservators have restricted their ability to communicate and to express their enthusiasm for their work by adopting too scientific a language.

> *Conservators must learn to communicate with other museologists. The urgent need for the members of this young highly technical profession to share information has led them to adopt the terminology of science, which is singularly unsuited to the humanities ... Until we rethink our vocabulary and language, the closer association with other museologists which we all desire may continue to elude us.* (Ward 1986, 62)

Communication between conservator and colleagues (and conservators and public) might be improved if conservators felt able to talk confidently about their own particular skills, their extensive knowledge and understanding of objects and their pleasure in interpreting and conserving them (Foley 1999b). It is worth noting that this problem of language has also been discussed by some archaeologists who feel that the adoption of a neutral, impersonal academic style may be hindering communication about archaeology (eg Hodder 1989).

The isolated conservator

While some conservators have risen to senior management positions, many remain comparatively cut off from the academic and administrative concerns of their colleagues. One of the difficulties is that many conservators (particularly archaeological conservators) work in comparative isolation from both the disciplines with which they are ostensibly collaborating and interacting, and the administrative structure and corporate aims of the institutions within which they work. Their work requires minute attention and they tend to work slowly and carefully, often staring down a microscope for long periods of time. They are based in laboratories which may be physically separated from other parts of the institution and which may not be visited much by archaeologists or curators, or they may be working independently in small practices in the private sector. The lack of contact may be caused partly because the conservator is more often seen as providing a technical service than as a participant in the debate about the interpretation and use of the material s/he is working on.

Museums and other heritage institutions are having to face considerable changes in order to meet new challenges, and change is almost always difficult to deal with. Because, with notable exceptions, conservators have tended not to play a part in museum administration, they may not be involved in discussion of possible changes. The result is that they can be cut off from decision-making within the institution and may feel (or become) marginalized and defensive. In other words they may find it difficult to work as members of a team concerned with the wider issues of the museum. This kind of situation is, of course, found in other professions too.

Conservators have reacted to this isolation by looking for support from colleagues outside their particular employing institutions. They have worked to develop and gain recognition for conservation as a discipline in its own right and as a separate profession. This has been beneficial in that it has raised technical standards of work, but it has had the unfortunate effect of causing many conservators to look inward (ie at their discipline) rather than outward (ie to consult and involve other disciplines) so it has exacerbated the isolation, and many continue to have a poor understanding of the contexts in which they are working.

There is even isolation within the conservation community. Not only has there been the long-standing distinction between the conservation of moveable and of immoveable artefacts but increasing specialization and expertise in individual areas of conservation (eg textiles, metals, buildings) has meant that specialist groups tend to communicate within their membership and there is little incentive to share information and viewpoint, and to develop policy across the whole discipline. This isolates specialisms from each other even where there may be common ground (eg paper and textiles, or different types of painted surface) and they may fail to learn from each other, or stimulate new approaches.

Change in the way conservators work

The role and function of objects conservators has changed. Traditionally conservators repaired and restored individual objects in the workshop, whereas now they are likely to be developing preventive care strategies for whole collections. The move away from a concentration on individual objects has been influenced by a number of factors: the increased understanding of deterioration and thus the adoption of preventive measures such as controlling storage and display conditions; the developing conviction that collections which are entities in their own right should be conserved as a whole rather than by singling out individual objects for attention; an increased caution in applying remedial treatments partly because it was realized that reversibility is not achievable in practice, and partly because the considerable tightening of the laboratory health and safety regulations has restricted the freedom to use some chemicals and types of treatments.

This change of emphasis from remedial to preventive has meant that conservation has been taken out of the laboratory and workshop and into the storage and display areas. Conservators have been able to become proactive rather than simply reactive, and to embrace the concept of collections management. Surveys of the condition of collections give conservators and curators a better picture of the state of the collections as a whole, and a tool for planning preventive conservation programmes. In terms of overall conservation, organized systems for preventive care of whole collections are more effective than piecemeal remedial treatments of individual objects (Keene 1996). From the conservator's point of view this has had both advantages and disadvantages. It has brought conservators into greater contact with other staff, and given them a more audible voice in policy-making, but it has downplayed the specialist knowledge and skills involved in remedial conservation, and even made the full-time presence of a conservator appear unnecessary. However, the development of collections management posts has given conservators opportunities of increased status and wider responsibilities, since many collections managers have started their careers as conservators.

The move from remedial conservation with its emphasis on the material character of individual objects is echoed by the growing emphasis on the conceptual significance of some types of collection, particularly ethnographic and contemporary art. Conservators have had to accept that deterioration and change may be an intended feature of certain types of object, not one to be limited and managed, but to be accepted and worked with (Ramsay-Jolicoeur & Wainwright 1990, Clavir 1994).

Change in employment patterns

Although there has always been a significant group of conservators working independently, particularly caring for paintings and other artworks, until recently archaeological conservators and those dealing with the range of social history and other objects commonly held in museums were generally employed by institutions. Some were employed in large institutions, such as the British Museum, which have specialist conservation departments with a range of staff, and some in small institutions which may have employed just one 'generalist' conservator. Now, however, a significant number of museums, certainly in the UK, are shedding their in-house conservators in favour of employing independent conservators whenever they need their expertise, with the consequence that more and more conservators are working independently, and finding that they need new skills related to project and business management. This comparatively rapid change in employment pattern can be traced through articles in the UK *Museums Journal* (eg Fraser 1997, Nightingale & Thomas 1998, Whittington 1998, Windell & Nightingale 1999, Heywood 2000). While the shedding of conservation posts in museums can be seen as lack of appreciation of the value of in-house conservators, working freelance can give conservators a new independence and the opportunity to perfect new skills (though, of course, adjusting to the loss of security of an institutional job can be difficult). The increase in the number of independent conservators has reduced the barriers between independent and museum conservation, and has improved the quality of both, because employers have started to define standards required when commissioning work. Loss of in-house conservators has gone hand-in-hand with involving other museum staff in the preventive care of collections. Curators and other staff are now very knowledgeable about preventive conservation, and this too has raised overall standards.

Conservator and volunteer

Conservators are often uneasy about involving inexperienced or untrained people for fear of damage to the objects and structures they are responsible for, but this is surely to deny the range of work which is relevant to conservation and to undervalue the special skills gained through the combination of training and long experience by practitioners in other fields. Some volunteers can contribute essential skills (particularly, for example, in the use and maintenance of industrial machinery) which conservators do not themselves have (Leigh 1994). There should be an opportunity for the student or assistant conservator, who is partially trained in conservation practice or trained in some skills only, to work with a conservator. It is crucial to improving the public understanding of conservation, and to encouraging wider participation and access, that more people are involved in conservation. On the other hand, conservators are rightly concerned about the use of volunteers without specialist guidance or supervision, enlisted by curators of museums without conservation staff. The fear is that volunteers may inadvertently damage objects simply because they are not familiar with the behaviour of old, degraded or damaged objects, nor with the requirements for the long-term stability of added materials.

Conservators and curators: attitude to objects and their use

One of the main sources of conflict is the many different ways in which objects can be used by the range of professions or groups which work with them. Archaeologists, curators, materials scientists, photographers, illustrators, exhibition designers, education specialists, antique dealers: all work with objects, and each group has different objectives and consequently deals with objects in very different ways. At the heart of this difficulty is the apparent contradiction in the dual roles of the museum: to preserve collections, but at the same time to make them accessible to visitors.

> *The curator must ensure that he passes on his collection intact and unharmed to subsequent generations. At the same time he must amaze, stimulate and educate his public. There is no substitute for the original object; if our successors only know material from photographs we will have failed as curators, whether we be conservators or museum specialists.* (Wilson 1982, 67)

Access is, in fact, what conservation is all about. It is not about total protection but about making objects usable.

Conservator as advocate for the object

Why do conservators have the reputation for being defensive and for more often saying 'no' than 'yes' when consulted about the use of objects and collections? Museums have a legal or contractual obligation to care for the objects they own, or have borrowed (Palmer 1989), and conservators' codes of practice emphasize the unique nature of each object and the need to respect this, as well as to 'serve as an advocate for the preservation of cultural property' (AIC 1994, III). These put a considerable burden of responsibility on the conservator who is concerned to protect objects from possible causes of damage. However, conservators often feel that they have the responsibility for objects without the authority to ensure their welfare, and, perhaps because they feel they have no voice, they can appear to be inflexible and over-protective.

> *Conservators ... can sometimes become super-conscious of the deteriorative effects of usage, and adopt a 'holier than thou' attitude, in attempting to prevent change, while not recognising the needs of other museum professionals to carry out their duties.* (Hodkinson 1990, 61)

Among some conservators there is a danger of seeing objects as clients, who take priority over the needs and interests of the people involved in the protection and use of the heritage.

> *Our loyalty is not owed to our institutions, organisations or colleagues but rather to the unique and irreplaceable objects that embody our history, culture and aspirations.* (Merrill 1990, 170)

While this attitude encourages conscientious care of collections or monuments, it is at the same time unhealthy as it discourages conservators from achieving a balance between the care of objects and their use by others such as curators, or visitors (Drysdale 1988). Inflexibility tends to isolate the conservator and move his/her focus onto the object and its perceived needs, and away from the needs of colleagues, the institution and the public. Once this situation is reached conservators may find it very difficult to relax their protective role and reach compromises with colleagues over the care and use of objects and collections. They are seen as too perfectionist, as well as being 'the "light, humidity, packing police"' (Michalski 1990a, 49). Constantly saying 'no' may mean the conservator will no longer be believed. This will not promote the cause of conservation, nor, in the end, protect the objects themselves (Barclay 1990). According to Fraser (1997) this over-protective attitude was one of the reasons given for the shedding of 11 out of 14 conservation posts from Glasgow Museums and Galleries, leaving a very small core staff to deal with the care of major collections.

Conservators need to accept that it is not possible to give the same amount of conservation attention to every object; some form of agreed prioritization is necessary so that the 'best' objects, at least, are given adequate attention (de Guichen 1982). Among some conservators there seems still to be the attitude that every object must be given the same amount of care (again, perhaps, a product of the burden of responsibility for the welfare of objects). However, this is simply not realistic: not only are funds finite, but it is accepted that some objects are more important and more informative than others. This concept of graded importance was embraced in the UKIC Archaeology Section's 1982 document which proposed five levels of conservation for excavated artefacts, from minimal care to the 'full works' of investigation, remedial treatment and restoration. It was also tackled in The Netherlands when the Delta Plan was set up to deal with the poor condition of collections in state museums. Objects were classified according to four categories, from Category A, the highly significant, to Category D, which covered material which should not have been included in the collection in the first instance, or was now in such poor condition that it was considered useless. Preventive conservation was improved for all objects retained in the collections but remedial conservation was given only to the most important objects (Talley 1999).

Archaeological objects

For most archaeologists excavated objects are a means to an end, the interpretation of the site. Much of an archaeological conservator's work involves elucidation of objects in order to discover what they are and how they were made and used. Because of the huge volumes of material from modern excavations and the limited conservation resource it is essential that there is some prioritization or selection of the objects to be treated. However, the significance of an object may not become apparent until it has been studied by a conservator, hence the importance of the preliminary investigative stage of conservation.

Unless they are working in the field, archaeological conservators normally work on individual items or batches of objects which have been removed from their original context, and perhaps also separated from the rest of the assemblage. It is the site itself which is the focus of study so the archaeologist is interested in the object as a tool for interpreting the site (and for publication), after which the collection of objects and other primary materials is deposited in the archive and rarely researched further as an assemblage. The excavator is effectively no longer interested in the material, having completed his/her work on it, so although some researchers may look at individual examples of object types from more than one site, much of the archive is no longer used. This means that there is little communication between curator, archaeologist and conservator about the material, and, given the large and growing size of many archaeological archives (and the conservators' many other responsibilities), it is difficult for conservators to monitor conditions effectively.

The curator's and conservator's attitude to objects

Between curator and conservator there can be a tension over the use and care of objects. The scholar can see the collections as 'hers' or 'his': s/he knows more about them than anyone else and so should have the authority to decide what is done with them, The curator, who is enthusiastic about working with the public, sees objects as available for a multiplicity of uses which may involve demonstration, handling and use by 'unqualified' people. Each type of curator may see the conservator's concern for the protection of the collections as an intrusion or an obstacle, but just as conservators need to relax their anxiety about the collections 'curators have to give up their "ownership" of their part of the collection. The collection does not belong to them; it belongs to the museum as a whole...The curator is only one user of the collection' (Spalding 1999, 33).

The cultural value of the collections depends on what is known about them as well as whether they are in good condition (Cossons 1992, Cossons 1994). Although knowing about the collections is traditionally the preserve of the curator, the investigative and diagnostic work of the conservator can yield significant information about an object, relevant not just to the curator but to the disciplines to which the objects relate, eg archaeology, cultural studies etc.

> *A conservator may be able to identify how the work was originally fabricated, what colour of ground was used, what transformation may have occurred in the transparency of the paint; whether the support has been cut down, whether signatures have been added or removed, whether whole elements have been over-painted or hidden by dirt or varnish; or indeed whether the materials used in the fabrication of the work were available at the time the work was supposed to have been painted. (Hill 1990, 20)*

The shift in museums from being firmly focused on objects to being also concerned with the interests of the public means that both curator and conservator must broaden their view of how objects can be interpreted and used, particularly if they are to affect policy decisions. Both curator and conservator must use their knowledge of objects to transmit some of their fascination to the public, and to involve the public in using and enjoying the objects in new ways because the collections 'are the means by which we communicate to our public. They are not an end in themselves' (Spalding 1999, 28).

Conservator and conservation scientist

If the material aspects of conservation practice are to advance still further in the twenty-first century, conservators and scientists need to work together more effectively. Conservation scientists are concerned with aiding conservators' investigation of objects, supporting conservation practice and developing new conservation processes. They can be effective only if they understand the problems posed by deteriorating objects, the constraints of current treatments, and if they have a clear objective: to improve the practice or effectiveness of conservation.

If scientists work in isolation with no regular contact with conservation and conservators their work will become 'at best, unhelpful, at worst condescending' (Tennent 1997, 18), to which could be added that if conservators work in isolation they will tend to continue to use the methods and techniques with which they are familiar. What is needed is better communication and more collaborative work between conservator as practitioner and scientist as researcher (Price 1992, Bradley 1997a, McCrady 1997). Progress may be hindered, too, by lack of collaboration between scientist and scientist, each a specialist in a different field, who, working together, could make greater impact on conservation (de Guichen 1991, Koestler 1994, Price 1996). A further problem may be that much scientific research takes the form of individually-funded discrete projects, so there may be little continuity, or building onto earlier research, because each project has to be 'new' in order to attract funds.

This situation is unhealthy for conservation, and is due almost entirely to poor communication where the conservator is as much at fault as the scientist. Van Asperen de Boer (1998) pointed out that there is the further problem of both archaeologists and art historians being wary of or even uninterested in the results of scientific work, an attitude which hinders their communication and collaboration with both conservators and scientists.

Language

The lack of a common language can be particularly acute between conservators and conservation scientists, often reflecting the different educational traditions of the arts and the sciences. The language of science may be foreign to curators or even to student conservators (Lasko & Lodewijks 1982, Brooks & Fairbrass 1997, Roy 1997). A similar language problem was identified by Vandiver and Wheeler (1991) when discussing the use of specialist terms by conservation scientists, materials scientists and conservators. Some conservation scientists describe their role as one of interpretation of science to the conservator (Bradley 1997b, Tennent 1997), which implies that the scientist understands conservation well enough to interpret the relevant science. One of the problems, though, is that neither side may understand enough of the other's concerns (and language). After all, a good interpretation requires the interpreter to be proficient in *both* languages, including the appropriate idiom and technical terms. Most conservators now do have some scientific training but few conservation scientists have a training in conservation, though they may learn much while on the job (de Guichen 1991, Price 1996, Bradley 1997a).

Different perceptions

The scientist and the conservator may have very different perceptions of the problem being studied. Podany and Scott describe a familiar situation:

> *Sitting side by side, the two may be looking at the same object and may be working towards the same objective, but the lens is often cut and polished quite differently. Indeed the very way in which each first approaches the problem at hand can be essentially disparate. Those differences are not sufficiently appreciated and both the conservator and scientist too often fail to look up from their particular view and glance in the other's direction, where both inspiration and common support can be found. (Podany & Scott 1997, 211)*

Effective collaboration relies on each participant understanding the other's situation, responsibilities and methods of work, and respecting each other's knowledge and skills. Conservators have to work with a range of variables within the object and within its environment, which provide multifaceted conditions which are difficult to describe and seldom repeatable, whereas scientists aim to limit the number of variables in order to obtain reliable and measurable results. Conservators want to see research which replicates reality, and are interested in the ease of application of any new treatment, as well as long-term results. Scientists cannot test as yet untried conservation processes on museum objects so have to work with modern materials made to simulate one or more of the problems presented by an object. Artificial or accelerated ageing has its limitations in terms of judging the likely success of a process, thus the long-term effects of treatments applied to objects can be evaluated only after considerable periods of time (Torraca 1996, Shashoua 1997).

The scientist needs to know enough about conservation not only to understand its aims but to respect the skills of the conservator, even if they seem 'unscientific'. Where it is impossible to measure, the conservator often relies on instinct guided by long experience of examining and handling objects.

> *Do not underestimate the conservator ... by feeling the properties of materials with very accurate instruments (his eyes and hands), he can cut his way through a multivariable problem more efficiently than the scientist, who is accustomed to proceeding by logical steps, and may have trouble identifying which variable is the relevant one. (Torraca 1996, 444)*

Publishing conservation

There has been very little publication of conservation treatments or discoveries in the generalist archaeological or museum literature and this illustrates the failures of collaboration or information-sharing between colleagues working in broadly similar fields. There are also comparatively few texts on conservation aimed at non-conservation colleagues. As conservation is not necessarily seen as making a direct contribution to the interpretation of an archaeological site, or to the way in which museum collections can be used by scholars and the public, the results of conservation work are not often thought appropriate for publication in archaeological, or museum journals: its contribution therefore remains largely unknown. A similar situation exists in the field of fine art where art historians tend to publish their work in the form of monographs whereas scientists and conservators submit papers on their work on paintings to specialist journals which the art historian would not normally read (van Asperen de Boer 1998). Conservation is not viewed as a suitable subject for publication in the 'hard' science journals with the result is that it is published in a narrow range of conservation journals which remain comparatively poorly known outside the profession (Price 1996).

Collaboration and teamwork, and crossing boundaries

Developing a common voice

It has been acknowledged for some time that there is a need for a common voice to put conservation on museum and public agendas (de Guichen 1982). This need is now being more widely recognized and constructive action has been taken with the formation of bodies such as the National Council for Conservator-Restorers (NCC-R, formerly the Conservation Forum) which has brought together a number of separate conservation bodies in the UK to work towards common objectives, including professional accreditation; and the European Confederation of Conservator-Restorers Organizations (ECCO) which has similar aims for the whole of Europe.

Collaboration in museums

Communication and collaboration, in the form of teamwork, are obviously important if conservation is to continue to develop and to be fully integrated in field archaeology and museums. All museum staff need to look across professional boundaries rather than look inwards, and to discuss problems and reach compromises, for the good of the profession and for the impact of the museum on the outside world:

> *Team working to secure sensible museum outcomes is easily dislocated if curators and conservators become territorial and look inwards to their own practice rather than outwards to the needs of the public and in their own disciplines. It is all too easy to fall into prescriptive approach because it is comfortable and avoids hard decisions and compromise: this object is in a museum therefore it is of equal importance to all the others ... Instead what is required in such a situation is a creative dialogue about the significance of collections, some calculation of the risk attached to different strategies for them, and what will be the effects of the museum on the wider environment.* (Hebditch 1998, 10)

In fact there are already examples of teamwork which bring together groups of different professionals. The benefits of collaboration involve not only more effective outcomes for the work being undertaken, but a gain in mutual understanding. Every time groups work together each person involved must gain a clearer picture of the objectives of the others (and of the constraints within which they have to operate).

Collections management

Conservators' participation in collections management not only puts them in contact with other staff, but brings together responsibility and authority and enables conservators to take the initiative rather than remain largely reactive. This is well illustrated by Keene's work at the Museum of London. Following the 1988 Auditor General's report on the poor condition of national collections in the UK, she decided to set up a working party to develop a methodology for condition survey at the Museum of London. The project not only influenced the management of collections at the museum but showed clearly the advantages to be gained from a good knowledge of the state of the collections, in terms of setting priorities and negotiating resources for conservation. She demonstrated that, with this information, it was possible to manage change by identifying causes of deterioration, diagnosing and then modifying trends in change, and assessing both the resources needed and the benefits ensuing from this approach. With this knowledge of collections, conservators become managers rather than being simply responsive to unanticipated change (Keene 1996).

Collections management also provides excellent examples of situations where conservation, curatorial, administrative and warding staff work together to develop policy or to improve practice. Disaster planning, development of integrated pest-management policies, risk assessment exercises and cost-benefit analyses all involve teams of museum professionals working together in order to achieve an appropriate outcome (eg Cackett 1992, Waller 1994, Wallis 1997, Cassar 1998, Ashley-Smith 1999a). Bringing everyone together means not only that all relevant factors are considered when decisions are taken, but each individual learns how the others contribute to achieving the objectives (Fig 49). The exercise undertaken by English Heritage in order to decide how to treat the eighteenth-century Chiswick tables involved the Head of Collections Conservation, the Director of Museums and Collections, a paintings conservator, a curator, a financial manager and the site manager. The costs and the benefits of each conservation/restoration option were considered and this ensured that the final decision was not taken on the grounds of cost alone and was 'owned' by all the stakeholders (Cassar 1998). A broadly similar group of professionals from Birmingham Museums and Galleries undertook the analysis for the treatment of the Wroxton Abbey state bed, and the process 'helped the team to own the final decisions on what to do with the bed. It was in this respect an excellent team-building exercise' (Arthur 1999, 4).

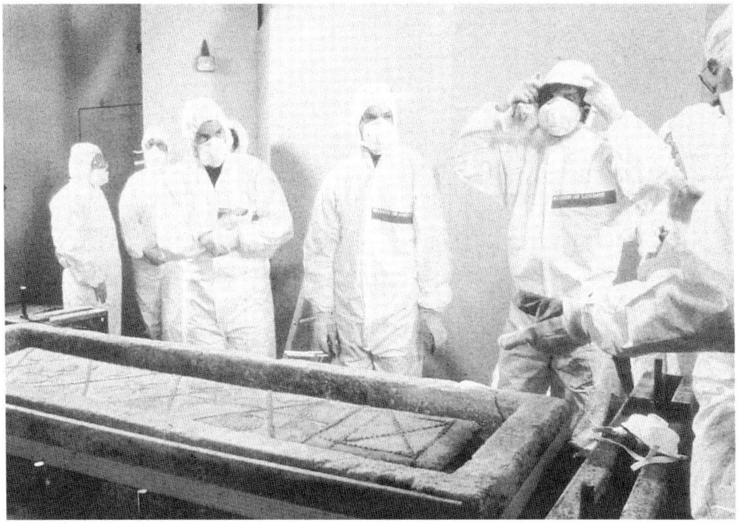

Figure 49 Teamwork: museum staff, including conservators, environmental archaeologists, technicians and object handlers, ready to open a Roman lead sarcophagus from Spitalfields, London (Courtesy of Museum of London).

Collaboration between specialists

Work on individual objects or conservation projects has the potential to bring different specialists together. Conferences and published papers enable conservators to report on collaborative projects where the input of colleagues such as curators, scientists or conservators working in other fields has been illuminating. However, the benefits of collaboration are seldom emphasized enough, nor are there enough joint papers which illustrate the contribution and the viewpoint of each collaborator (but see Podany & Scott 1997).

Deciding on what is significant about an object is essential before conservation can proceed, and for large and complex projects conservators and curators (or owners) must work together in establishing what is important. A good example of this form of collaboration was the treatment of Whistler's Peacock Room, now in the Freer Gallery of Art in Washington DC. The art-historical and scientific research had an important influence on the approach to conservation, and its success in revealing and maintaining the visual impact and thus the significance of the room (Samet *et al* 1990). Another good example of collaboration was the work undertaken at the British Museum on a seventeenth-century watch that had been retrieved from the River Thames in London. Here two conservators, each working in a different specialism (archaeological and horological conservation), worked with a scientist to decide how best to approach the corroded watch, and each conservator clearly learned from the other during the process (Meehan *et al* 1996).

Collaboration between conservators could have other practical benefits. With more conservators working independently, fewer have direct access to sophisticated equipment and other facilities, including libraries and subscriptions to journals normally supported by institutions. There are signs that independent conservators are considering joining together to share premises and facilities. Formation of partnerships of this kind will make it possible to afford not only expensive equipment, but, for example, secretarial support. This echoes similar arrangements made by medical doctors and lawyers, which are apparently very effective, and should enhance standards of work, as well as making good business sense.

Cross-professional meetings and conferences

Bringing together different members of the heritage professions at meetings and conferences can result in wide-ranging discussion and the presentation of stimulating and innovative papers. It is a pity that the pressure to concentrate on and extend individual specialisms means that this cross-fertilization does not happen often enough.

Examples of this kind of meeting include the seminar involving curators and conservators at the National Gallery of Canada in 1989, at which the challenges of conserving art, including modern and conceptual art, were discussed (Ramsay-Jolicoeur & Wainwright 1990); the workshops held at Dahlem, Germany, in 1992 and 1996, which brought together conservators, scientists and curators who tackled such questions as change in artefacts and the value of cultural heritage (Krumbein *et al* 1994, Baer & Snethlage 1997); and the series of symposia entitled 'Material Issues in Art and Archaeology' which has been held in the USA every two years since 1988, bringing together materials scientists, conservators and archaeologists and producing a very wide range of papers on the nature and behaviour of materials and objects (eg Vandiver *et al* 1997).

The benefits of different conservation specialisms meeting to compare philosophy and practice can be seen in the papers arising from conferences such as *Paper and Textiles: The Common Ground* (Butterfield & Eaton 1991) and *Lining and Backing* (Durham 1995). These conferences brought together conservators working on paper, textiles and paintings and enabled them to compare and learn from the different approaches in their different specialisms. Encouraging conservators from different specialisms to meet informally and share views is also vital, as at the round-table meeting organized by English Heritage in 1998 which covered topics such as the meaning of objects, the use of information technology to communicate conservation, and the mission of conservation (Foley 1999b). More meetings which produce this kind of cross-fertilization would be beneficial to the whole discipline.

Unfortunately, the results of many of these wide-ranging meetings are published in stand-alone volumes or series, rather than in more widely circulated journals, which means that they do not reach a large enough audience even within individual specialisms.

Benefits of collaboration

Good working relations and cross-fertilization arise out of formal or informal collaboration, out of taking an interest in other specialisms or disciplines, and looking beyond the confines of one's own responsibilities. Conservators could take a greater lead here by attending meetings of, or even joining other professional bodies outside their immediate specialism, and by seeking and publicizing the benefits of collaboration. There is no need for them to feel isolated, or to feel they have no voice.

9 Development of conservation: professionalism, training, research

Introduction

The aims of this chapter are to explore the issues of professionalism, training and research, all of which influence standards, and effectiveness, within the practice of conservation. Development of a profession of conservation and improvement in training will affect attitudes to the role of conservation within museums and archaeology and to the development of conservation in the future. Development of scientific research affects conservators' ability to implement improvements to treatment.

In 1983, at the opening of a new building for the Conservation Centre at New York University, Baer (1989) suggested that there were several questions which needed to be asked, including whether conservation had become a profession, whether it was recognized in the community, whether conservation was producing professionals with a combination of scientific, art-historical and practical knowledge and skills, whether training covered all types of cultural material, whether training in an academic setting was appropriate and whether training was able to meet the needs of both the developed and the developing world. These remain relevant questions over twenty years later.

Is conservation a profession?

The evolution of conservation is seen by most practitioners as encompassing not only development of practice, but the growth of a recognizable discipline, improvement of standards of training, and the emergence of a profession of conservation. The earliest professions included law, medicine and the church, but many other occupations have joined the ranks of the professions such as architecture, engineering and, now, 'the museum professions' (Kavanagh 1994). The questions which arise are: is conservation a profession, if so what are the effects of being a profession, and is it to the advantage of care of the heritage that conservation should be a profession?

The term 'professional' is often used loosely. Attempts to define a professional are widely varied but some characteristics seem to be common to most descriptions. A professional possesses specialist knowledge not shared by the wider community, and his/her authority relies on expertise which may be publicly recognized in some way (sometimes formally recognized by the state). S/he offers some form of service to the public, and is a member of an organization of peers which sets and controls standards of behaviour, including the requirement to adhere to a code of practice or ethics (Koehn 1994, DiMaggio 1994). Another characteristic which has been added recently is the commitment to keep his/her expertise up to date (which probably reflects the increased pace of change in the practice of the professions).

However, it is apparent that there is considerable mistrust of professions since they are, by their very nature, exclusive (George Bernard Shaw is reputed to have described professions as 'conspiracies against the laity'). Whereas the advantage of the professional system was perceived to be that the public could expect expertise coupled with certain standards of behaviour, such as not to be taken advantage of, or subjected to injury (eg through medical or architectural malpractice), there is now some mistrust of professions, and less confidence that the system functions effectively on behalf of the public it supposedly serves. Thus professions, such as medicine, are no longer seen as reliably authoritative, and are increasingly having to open themselves to public scrutiny and to become answerable for their actions. Another potential reason for tension or mistrust is that the professional is expected to adhere to the standards set by the (external) professional association as well as being responsible to his/her employer. 'His work, its standards, its goals, its vision, are set by the standards, the goals, the vision of a profession ... in other words they are determined outside the enterprise', ie outside the museum (Drucker, quoted by Hatton 1994, 164).

At the beginning of the twenty-first century conservation is going through a phase of self-examination and self-definition. Both internal and external pressures appear to be encouraging the formalization of a profession. Within the discipline conservators are concerned about status and recognition, and about changes in employment (eg fewer jobs in institutions), and externally employers are interested in standards of practice (as reflected, for example, in the establishment of the Conservation Register by the UK Museums and Galleries Commission, which has now been passed to the conservation profession to administer).

So does it matter if conservation is or is not a profession? In the hands of the inexperienced or the unscrupulous, conservation can undoubtedly cause damage to the cultural heritage, so one advantage of conservation's becoming a profession and being subject to control is that damage should be limited. Another advantage is that members of a conservation profession would adhere to a common statement of its mission (which would clarify what conservation is), and their practice would be defined and controlled by a common code of ethics.

A profession would ensure that its members were practising at an acceptable level, and adhering to the code of ethics. Hitherto many or most conservators were employed by institutions and so were, in a sense, vouched for by their institution (the label of 'conservator at the British Museum', or 'at the Smithsonian', implies certain standards of practice). At present anyone can call him/herself a conservator whether s/he has any expertise or not, and with more and more conservators working independently the lack of recognized standards presents a problem to the employer, and potentially endangers the cultural heritage. There have been situations where conservators have advertised their services and hoped to indicate a certain standard of practice by describing themselves as being a member of a particular conservation body. However, with a very few exceptions, most existing conservation bodies do not (yet) vet the practice of their members, most exist to promote the discipline of conservation, so this can be misleading (but it indicates the need that many conservators feel to be able to say they have reached recognized standards of practice).

A profession would not only set and monitor standards of practice but would also act to protect its members' interests. So it would attempt to improve pay and conditions of employment, it would protest about shedding of conservation posts by museums, or support refresher training courses for its members. Thus becoming a regulated profession would have advantages for the preservation of the heritage, for employers, and for conservators themselves.

To what extent is conservation already a profession? Conservation is now acknowledged, by conservators, to depend on both a particular range and depth of knowledge, and specialist expertise, without which it is not possible to practice effectively. This is bound up with the evolution of the discipline of conservation, and is what distinguishes modern conservation from the old 'craft' of the repairer/restorer. However, this expertise is not yet widely acknowledged outside conservation, and it does not yet have wide public (or state) recognition. Because the practice of conservation is not widely understood it is not yet acknowledged to provide a useful public service (in contrast to the probable general agreement that society needs some form of medical profession and legal profession).

To reinforce this lack of public support, there are intermittent, and sometimes very public, attacks on what are seen by some as the damaging practices of restoration (ArtWatch International 1999) which indicate that, among some of the public at least, conservators and restorers are certainly not seen as 'admirable' experts, and may be seen as causing damage to the cultural heritage. Conservation is fragmented into a range of different conservation bodies, few of which control their standards of practice. Many have codes of practice but few have any system for 'policing' them, and few insist on members keeping up to date with their subject or methods. This fragmentation also means there is no clear voice promoting conservation to the public, so conservation does not apparently behave like a profession. On this analysis, lack of public support is a considerable obstacle to the development and recognition of a profession.

What are conservators doing about this? In fact, at the beginning of the twenty-first century, there is considerable evidence that conservation is moving towards becoming a profession. The body of knowledge needed for practice is characterized by a strong theoretical and philosophical element which is developing in parallel with the increasingly specialized nature of information underpinning the technologies of practice. Knowledge is being extended and disseminated through research and publication, and codes of ethics are being revised and added to in order to cover changing circumstances.

Most importantly, fragmented conservation associations are joining in the development of accreditation schemes for their members, eg NCC-R (the National Council for Conservator-Restorers) in the UK which embraces a number of different conservation bodies, ECCO (the European Confederation of Conservator-Restorers' Organizations) which is attempting to harmonize professional standards across Europe, and AIC (the American Institute for Conservation) which represents a wide range of conservators in the USA. Coming together in this way, particularly to set up accreditation (certification) schemes, requires that conservators of different specialisms join in clarifying, and agreeing, their role and function. A unified view embracing a clear mission and set of standards must in its turn have some impact on public perception. If conservators are clear about their role they will be able to transmit a clearer message to employers and the public. This in its turn will strengthen the profession's arguments for formal state recognition, either of the professional body (as would be required in the UK) or of the individual professional through his/her mode of training (as in much of the rest of Europe, see Foley & Scholten 1998).

The establishment of a profession should be beneficial to the preservation of the cultural heritage in that better and more consistent standards of practice, together with improved public understanding of conservation as a distinct specialism, should prevent damage. It should be beneficial to the development of conservation in that knowledge, skills and ethics will be continually reviewed and extended, and to conservators in that accreditation will indicate professional competence. But the profession must be used responsibly, to look outwards and to work constructively with employers and with other groups such as archaeologists and curators, and different interest groups within the public, including amateurs and volunteers. It must not be allowed to become a 'closed shop'. If the cultural heritage can be seen, in some senses, as belonging to everyone then a conservation profession must be answerable to the public, particularly as conservation has the power to destroy as well as to preserve. A new conservation profession should declare itself as open and answerable from the start (it should learn from the example of the old professions, such as medicine which, in the UK at least, is losing public support because it is clinging to its old autonomy and appears reluctant to make the transition to becoming open and answerable).

The establishment of a profession has implications for the perceived status of conservators among their colleagues, and for their education and training. Professions often implement accreditation (certification) of their members through some form of control over training programmes, which then allows approved programmes to provide a route through at least part of the accreditation process. This implies not only the vetting of the content of a training programme but also regular monitoring to ensure that the content and standards remain acceptable to the professional body. Accreditation is, itself, no longer a 'one-off' process since the professional is required to keep his/her knowledge and skills up to date through a process of continuing professional development (CPD). Once accreditation systems are established the conservation profession may wish to review training programmes. Here, too, it should adopt a broad view, and it should be open to different forms and emphases of training rather than becoming restrictive. Welcoming a range of different approaches to training will be healthy for the development of the discipline as well as the profession.

Developing conservation education and training

Background

Training in conservation has been and remains very varied, partly because of the different traditions which have contributed to the development of conservation and partly because of the growth of specialisms. Some common objectives are needed to enable conservation training to promote coherence in the discipline, but at the same time it is important that the future career path of each student is considered, and that there is effective training for each of the different types of conservator now emerging (eg conservation technician, preventive conservator, conservator/ restorer, consultant conservator etc).

The varying standards of practice within conservation arise, to some extent, from the range of different types of training available, from one-week courses to postgraduate degree programmes. They also arise from the fact that there are still, at the beginning of the twenty-first century, some conservators in practice who may have had little training, and some specialisms for which there is little training available. However, the design and development of modern training draws on educational thought and offers an opportunity for improvement of conservation as a whole.

The training of the craft restorer

In the past craft restorers learned their skills through some form of apprenticeship (and some restorers still do). Apprentices were taken on (often individually and always in small numbers) to work with experienced restorers or conservators who taught them the techniques and skills of restoration/conservation. 'Apprenticeships were rather loose arrangements between practising restorers/conservators and the student who often learned through an osmosis system' (Bacon & Brown 1996). Apprentices were taught to practice in the way the 'master' did, they learned the skills that the s/he had perfected and they were expected to carry on that tradition. In other words apprenticeship tended to be inward-looking with learners developing within a more or less closed system.

Apprentices tended to learn through the experience gained from whatever projects passed through the studio/workshop. This system could provide excellent training in the required skills but much rested on the particular qualities of the supervisor and it required both individual attention and plenty of time to ensure that all aspects of practice were covered. A weakness of the system was that an apprentice could learn the practical techniques and skills without acquiring any understanding of why they were used or how they worked. 'Those who have trained on the job have sometimes done so, it is said, by repetitiously copying their seniors, often acquiring manual skill at the expense of understanding. They are able to act but not to analyse' (Foley 1992, 27).

Development of formal training in conservation

As conservation developed as a discipline during the twentieth century, so did a concern with how conservators should be trained. Gradually formal training developed, such as courses at the University of London, Institute of Archaeology (Fig 50) (Hodges 1963, Gedye & Hodges 1964). At the first conference organized by the International Institute for Conservation in Rome in 1961 a section was devoted to 'Education and training of conservators and restorers' (Thomson 1963). After its foundation in 1959, the International Centre for Conservation in Rome (now ICCROM) prompted further discussion (see ICOM 1969, and National Trust for Historic Preservation 1976), which helped to define training in general and the Centre's role in training in particular. By the last quarter of the twentieth century there was a wide range of different formal conservation courses, many based in universities. ICCROM provided a programme of international courses (Stanley-Price 1989), and there was an active Working Group on Training in Conservation and Restoration of the ICOM Conservation Committee. Even so, surveys of provision indicated that there were still untrained people working in conservation: for example as late as 1989 a UKIC survey revealed that only 61% of people entering the profession in the UK had conservation qualifications, though in part this reflected gaps in provision, eg lack of training in the conservation of furniture or natural history collections, rather than failure to take advantage of relevant training (Corfield *et al* 1989)

Figure 50 The old teaching laboratory at UCL Institute of Archaeology. In use since 1958, this laboratory has seen generations of students now practising as conservators world-wide. In 2000 this lab was redesigned and refitted (Courtesy of UCL Institute of Archaeology).

Academic education

Many of these new courses were based in universities, but was provision of conservation education in universities an improvement on the old apprenticeship system? The difficulties of conservation training were encapsulated by Stanley-Price in 1990.

> *Conservation, in common with other areas of expertise, has gradually moved away from craft-based apprenticeship training to academically-based formal programmes of education, appropriating in the process the prestige of a university degree. There is little agreement, however, on what status is conferred by an academic degree in conservation, or in other words, what is meant by a qualified conservator? (Stanley-Price 1990a, v)*

Academic education provides intellectual skills. For the potential conservator an education in art history or archaeology is an excellent foundation, and means s/he understands the significance of the objects and their role in the study of the particular discipline. An education in conservation should provide an understanding of the way materials deteriorate and the range of treatments available, but education alone does not train students to practise as conservators since it does not equip them with the practical skills. Some senior practitioners who had trained through apprenticeship were very concerned that 'academic' conservators would simply not have the manual skills.

Balance of science and humanities training

As the conservation discipline developed an increasingly scientific base, many academic training programmes incorporated teaching on general scientific concepts and on the chemistry of materials. There was no clear agreement about how much science or which aspects of it should be included, and as recently as 1986 a meeting of the ICOM Committee for Conservation Training Working Group was held at the British Museum to exchange views on this topic (Pearson & Winsor 1986, Feller 1989).

Thus not only was there the issue of how academic a conservation training should be, but also how scientific. A major concern in some fields of conservation was that emphasis on science would leave too little opportunity for education in the humanities and in understanding the significance, or 'otherness' of objects (Talley 1990). There was concern, too, that without adequate knowledge of archaeology, art history and related subjects conservators would not communicate effectively with curators and other future colleagues.

Training in practical skills

Many conservation training courses (including those based in academic institutions) had retained some aspects of the apprenticeship approach to practical training. This is, perhaps, not surprising, since when the first courses were introduced it was an established method of teaching practical skills. (Other forms of university teaching which are still in use, such as lectures, have an equally long tradition.)

On his first encounter with conservators at a meeting organized by ICCROM in 1980, a University of London education specialist observed, with apparent surprise:

> *I noted that there was a tendency for some conservators to adopt approaches to teaching which were based upon the Renaissance apprenticeship model. This was interesting while a*

small number of people were undergoing training, though it was not of much use to those who had to run degree courses. (Ferguson 1992, 412)

In the context of a formal training course, an apprenticeship approach has some serious weaknesses: it is not effective if the time available is limited and the number of students comparatively high, it cannot guarantee to expose each student to the range of experience he/she needs, nor does it necessarily encourage the wide-ranging thought and judgement needed to practise at a high level. For many students it is difficult to make an effective connection between the practical and the academic parts of a course. The particularly able and motivated student will surmount some of these difficulties but the average student can feel lost, and both may be discontented.

Those involved in conservation teaching were increasingly aware of these difficulties. On scanning the literature on training and education in conservation a gradual trend can be seen, starting with discussion of the knowledge and skills which are needed (eg Coremans 1969), then moving to exchanging information about the structure and content of different conservation courses (eg Hodges 1963), until, by the last twenty years of the twentieth century, conservation teachers were beginning to take account of, and test, educational theory (ie to look outside their own discipline). This has much to do with the initiative of ICCROM and the Getty Conservation Institute in involving education specialists in the discussion of conservation training and in the implementation of a wide range of different 'train the trainers' courses.

Current debate on conservation training

Discussion of conservation training during the last years of the twentieth century was influenced by the debate about whether conservation was, or should be, a profession. It was also affected by the wide-ranging debate within educational and professional circles about vocational and professional training.

Discussion of preparation for a vocation or a profession centred on the relative contributions of education and training. If education is defined as providing knowledge and understanding (sometimes expressed as 'knowing that'), an educated person will know about relevant facts, concepts, principles and procedures and will be capable of using this knowledge to analyse, solve problems and reach decisions.

Education is about encouraging people to ask questions and to seek answers. It is also about being able and willing to reflect upon what one does and why one does it. To be educated is to be productively ill at ease. (Ferguson 1992, 413)

If training, by comparison, is seen to provide practical knowledge and practical skills (sometimes expressed as 'knowing how'), a trained person will know how to do things, how things work and what happens when something is done.

Training involves routine, repetition, discipline ... Well-trained people know how to follow the rules, how to perform skills in particular ways. (Ferguson 1992, 414)

It is immediately apparent that the old craft restorer was usually trained but may not have had much education, whereas the modern conservator needs both education and training. However, the character of education and training seem to be essentially opposed: one has few boundaries and requires wide-ranging thought, while the other has definite boundaries and does not expect variation of the established processes. A crucial difficulty with preparing conservators for practice is to link education and training so that their practice is informed and influenced by knowledge and understanding.

In the last years of the twentieth century discussion of conservation education and training focused on the difficulties of providing a balanced programme which covered both the academic and the practical, and provided students with a sufficiently scientific approach, while giving them awareness of the humanistic aspects of studying and appreciating objects. By the end of the century conservation teachers were concerned with the requirement to prepare their students for practice at a high level (professional practice), to enable them to identify and solve problems, to communicate with colleagues in other fields and with the public, to work effectively in teams, to undertake research, to prepare and submit proposals for jobs, and to run projects effectively (Leigh 1990). These concerns were not unique to conservation education but affected all heritage (and other) professions. For example, debate focused on whether problems such as poor communication in professions such as medicine were due to inadequate training.

Education theory provides some insight into how these different components might be effectively brought together. Eraut has defined three different types of knowledge needed for professional practice: personal knowledge, propositional knowledge and process knowledge (Eraut 1992). Personal knowledge relates to attitudes and experiences acquired before, or outside, training (and affects, for example, interpersonal skills); propositional knowledge covers the theories and concepts relating to a discipline (ie academic knowledge); process knowledge covers many of the attributes necessary for practice such as planning, problem solving, practical skills and communication.

A common difficulty with current conservation training is that students do not find it easy to link, or apply, the information received in the academic part of the course to the practical work, ie to link the propositional with the process knowledge (as defined by Eraut). Professional practice requires both types of knowledge, but effective process skills cannot be supplied simply through routine work on objects. Additional input is needed to enable students to develop the diagnostic, practical and study skills which will equip them to identify and solve the problems presented by 'real' objects.

Reflective practice

Similar difficulties have been identified in other spheres, such as medicine, and during the last quarter of the twentieth century new approaches were developed to provide education and training for practice. What is being aimed for in the professional practitioner is known as reflective practice, which is the ability to draw on and use a wide range of different types of knowledge and experience in making judgements and reaching decisions, and it is also the ability to be self-critical (Brockbank & McGill 1998). Effective education is partly 'about being able and willing to reflect upon what one does and why one does it' (Ferguson 1992, 413). Reflective practice is what is expected of all professionals whether they are lawyers, architects or conservators.

Diagnosis and problem-solving

Much of what professional practitioners do, including conservators, requires diagnostic and problem-solving skills, and the approach to education now used to prepare reflective practitioners, known as problem-based or independent learning, concentrates on developing these skills (Barrows & Tamblyn 1980, Schmidt 1983, Walker & Wright 1996, Taylor 1997). This approach presents students with a series of problems, each of which requires them to work independently (and/or in groups) and to draw on, and develop, their propositional and process knowledge.

Problem-based or independent learning encourages students to use a wide range of resources and to discuss and reflect on their own experience. It is essential that both teachers and students are well prepared because their respective roles are very different from what is normally expected, in that students take charge of their learning and the teacher becomes a facilitator (Taylor & Burgess 1995).

Inputs, outcomes and competence

Another subject of discussion in education circles generally (and in conservation) was the relative merits of evaluating education and training programmes by looking at the input, eg in the form of a syllabus, or by looking at the outcome, that is at how effectively the student could perform in the chosen occupation or profession after completing the programme. It is obviously quite possible for a student to sit through a series of excellent lectures and to attend a number of carefully structured training workshops, but not to be able to perform effectively in the particular activity at the end of the process. Thus there are strong arguments in favour of the evaluation of outcomes because this method tests whether a practitioner can actually do the job.

This effective performance is often described as competence. Competence in a given task is achieved through training and may aim for a single acceptable level of performance. Professional competence is achieved through an amalgamation of education, training and experience. The experience enables the professional to develop insight and the ability to use judgement and to emerge as a knowledgeable, skilled and reflective practitioner. In this case the assessment of competence must take account of the professional's ability to deal with a wide range of variability, and the complexities of competing factors. (For discussion of outcomes and competence see, for example Burke 1989, Jessup 1991, Atkins *et al* 1993, Barnett 1994.

The assessment of competence has become controversial because some of the assessment systems based on this concept are apparently mechanistic and seem to be suited to occupations which do not require a large knowledge base. When these same systems are applied to professions which require an extensive education there seems to be little scope for assessing the range and depth of knowledge required or the subtleties of problem-solving, and strategic planning. Nevertheless the underlying concept of assessing a person's ability to perform effectively in an occupation or profession is a useful one, since it should provide a better indicator of ability to practice than 'paper' qualifications (Cronyn & Foley 1996). Probably the first suggestion that this approach could be applied to conservation was made as long ago as 1984 (Seguel 1990). (For an example of a detailed system proposed for the museum field see MTI 1995, and for criticisms see Ferguson 1992, Talley 1992, Pearson 1999.)

Defining the activities of conservation

In the conservation world the second half of the twentieth century was characterized by attempts to define the elements of conservation, and these had a direct bearing on the evolution of the profession and of training.

Attempts to describe and define the activities, training and qualifications for conservation ranged from finding out how conservation was being provided to drawing up suggested 'standard' syllabuses, to analysing the functions of the conservator. Several surveys were undertaken during the last part of the twentieth century to find out about the provision of conservation and to develop a clearer picture of employment patterns, working practices, training and range of specialisms, including gaps in provision of training (eg NIC 1984a, Corfield *et al* 1989, Giles 1989, Ramer 1989, Winsor 1998).

Other attempts to define conservation involved producing syllabuses or curricula intended to provide a common guide to training. This was done both to set standards and to define the knowledge and skills required by a conservator. Implicit in these activities was the need for some kind of clear standard and coherence (Keck 1963, Coremans 1969, Fitch 1982, NIC 1984b, Cronyn & Pye 1988, NIC 1990, Pearson & Ferguson 1996). However, one of the weaknesses of the definition by syllabus is that the underlying philosophy may not be clearly articulated, nor is there any indication of what the student should be able to do on completion of the course (Stanley-Price 1990b).

Definitions of the conservator, or the functions involved in conservation, provide some aspects of what may be absent from a syllabus – an indication of what the conservator does and how s/he should behave. In 1984 the ICOM-Conservation Committee Working Group on Training in Conservation and Restoration defined the 'Conservator-Restorer', which was one of the first attempts to arrive at an internationally relevant description of the responsibilities of the conservator. The title of conservator-restorer was adopted to reflect the different usage in different languages (ICOM-CC 1984). Similar attempts to define the conservator were subsequently undertaken by Pearson (1990) and by ECCO (Wikström & Schießl 1996).

The starting point for assessing competence is to understand what the occupation consists of, what it is that the person has to know and to do to be considered competent. Arriving at this point involves functional analysis. The analysis of conservation undertaken in order to prepare the occupational standards for the Museum Training Institute in the UK involved deconstructing the activities, and the decision-making and planning processes, involved in conservation. It was a challenging and useful exercise because a team of conservators had to analyse exactly what constitutes conservation practice.

At the end of the process agreement was reached on what was, in effect, a definition of the occupation or profession in the form of a description of the activities common to a wide range of conservators in the UK. This was the starting point for the evolution of further generations of standards. Using these standards, further analysis was undertaken by Foley in 1998, in an attempt to arrive at professional standards which could be agreed to define the work of a conservator across the whole of Europe (Foley & Scholten 1998, Foley 1999a).

Another preoccupation was to define what was needed to confirm entry to the profession. The question asked was 'should professional recognition depend on an academic education as a fundamental requirement, or should it depend on the ability to practice effectively?' At the end of the twentieth century a further generation of the standards was developed by the National Council for Conservator-Restorers (NCC-R) for the accreditation of conservators in the UK. These standards had evolved from the Foley standards and were intended to assess professional competence, but have been designed to take account of both process and propositional knowledge. To do this they include both functional and professional criteria.

The professional criteria require demonstration of:

- professional values and management of value-conflicts
- intelligent practice and professional judgement
- reflection, enquiry and personal professional development
- sensitivity to the cultural context of materials and the values of people
- effective, appropriate and sensitive communication.

Thus the professional criteria are a basis for the assessment of a candidate's propositional knowledge, reflective skills and ability to make judgements, all the skills required by a professional practising at a high level. Taking both the functional and professional elements together, these standards should enable the profession to adopt a credible accreditation system. The standards are also useful for training courses in that they provide a good indication of the kinds of skills and abilities which they should encourage, so that their graduates have an effective foundation for eventual accreditation.

Preparation for the conservation profession of the twenty-first century

So what should be expected of conservation training for the twenty-first century? It is generally agreed that preparation should embrace an academic education, as well as a training in the required practical and managerial skills. It is also widely agreed that the training should draw on current educational theories now being applied to other types of professional training. Training also needs to take account of current (and future) trends in practice, and in the profession as a whole. This is particularly important now that so many conservators will work independently

For all students the starting point must be an academic education in the required science and archaeology and art history, which covers methods of research and analysis, and reflective practice, as well as educating them in the nature of the heritage and the theory and principles of conservation. Unless they know and understand the wider discipline within which they are working, and the significance of the material culture they are working on, they will not be effective conservators, nor will they work effectively with their colleagues.

Rather than relying on the old unquestioning apprenticeship system, practical training should draw on different approaches including the practice and perfection of specified skills, and the use of problem-based learning which encourages students to draw on their knowledge and their skills and develop a reflective mode of practice (Stanley-Price 1990b). Practical training should foster observational, investigative and aesthetic skills, and should be wide-ranging and flexible. No training should inculcate only one approach or method, otherwise graduates will be unable to discriminate when new approaches and methods become available.

> *their education should give students a firm conceptual armature around which to organise their professional activities when they enter the field. What they do with such conceptual equipment will depend partly upon their own interests and capacities and partly on how they apply them, in light of their widened comprehension of this growing and immensely important area of human activity. (Fitch 1982, xiv)*

As well as the skills particular to conservation, generic skills of communication, team-working, project design and management should prepare students for professional practice, and help to break down barriers between the various heritage professions. Graduate conservators should be confident and outward-looking, and have a good understanding of the institutional context (Leigh 1990). They may need new skills which equip them to communicate and work with local communities, or to mediate in contested situations (English Heritage 2000b).

It was also agreed that preparation should take account of the development of professional accreditation systems, in that some training programmes may be recognized eventually as fulfilling a stage in professional accreditation. Linked to this is the fact that a significant number of trainers and several professional bodies are interested in the assessment of outcomes or competences based on a definition of the functions of conservation.

Different routes

Students may achieve their conservation education and training in different ways. They may enter their formal conservation training having previously taken a relevant degree in, say, art history or archaeology, and possibly having had a period working as a volunteer in a museum. This prior experience or education can be very important in shaping students' interests and abilities before entering a conservation programme, and will affect their development during the programme. Assessment of outcomes enables credit to be given to similarly high levels of competence achieved through different routes (ie different combinations of training and experience). This encourages variety and will strengthen the profession in a way that reliance on a 'standardized' route would not.

Professional development

Students need a framework for the development of their practical and professional skills which will allow them to take advantage of different opportunities and to go on developing throughout their professional life.

Evidence of continuing professional development is a normal requirement for maintaining accredited professional status, thus the developmental framework should provide students with the skills of planning and developing their own career. The concept of continuing development used to demonstrate professional capability normally consists of two components: a plan which sets out intended career development, which might, for example, list new knowledge and skills that are aimed for and how these might be achieved, and a record of what has been done and what impact it has had on career development.

Work experience

It is clear that the discipline of conservation is now so wide-ranging in terms of knowledge and skill that it is impossible and inappropriate for conservation training institutions to cover everything, and that it is much more effective for early professional experience to be provided in the normal working context, ie museum or similar institution (Foley 1989, Holben Ellis 1990, Mathisen & Rushfeld 1999, Pye 1999).

However good the practical training in the academic context, the most effective way of developing professional skills is through workplace experience. This provides students with an understanding of the pressures and constraints of professional practice and of the requirements of the wider institutional context. Understanding the workings and priorities of an institution at first hand is particularly important for those students who may go into independent practice and be employed by institutions for short-term projects.

Extended internships have particular advantages in that students learn to work within a wider team, and to deal with the time, financial and other constraints imposed by the objectives of the institution. This kind of work experience is the equivalent of employment at a junior level, so gives the student the opportunity to demonstrate his/her competence in the workplace. This has great advantages for future career prospects, and recent research undertaken by the Institute for Employment Studies for the UK Museums and Galleries Commission has shown that internships provide a very effective route into full-time employment (Jagger & Aston 1999).

The advantage to the institution and the profession is that there is a regular influx of freshly educated and questioning conservators working together with established staff. The system also has the advantage of giving the profession a role in training future professionals, but it does require that individual conservators accept training provision as one of their professional responsibilities (Pye 1999).

The international perspective

The final question which Baer (1989) asked about training was whether it was able to meet the needs of both the developed and developing world, ie was it suitable for a multicultural audience? It is important that conservation trainers and future conservators respond to cultural diversity as the involvement of new and wider audiences is welcomed and encouraged (Hooper-Greenhill 1997, Federspeil 1999). This is important also because conservation is already an international activity, with many examples of international collaboration fostered particularly by ICCROM and by the Getty Conservation Institute, but typified by, for example, COOL (COnservation On Line) the website based in USA and widely used internationally, and by a mobile international workforce. The advantages of international training are emphasized by Stanley-Price (1989), but many of the long-established training programmes are based in the Western, developed, world and are inevitably influenced by the local social and professional contexts, as well as by factors such as the national administration of archaeology and museums. Teaching is biased towards local conservation problems and requirements, eg training in London tends to emphasize the problems of a temperate rather than a tropical climate. Training in the developed world may also fail to address the constraints faced by developing countries where much of the equipment and conservation materials relied on by conservators working in the West are simply not available. In these respects the long-established training programmes have significant weaknesses for students coming from other countries and cultures. However, provision of a developmental framework which encourages a student to take the foundation training and extend his/her knowledge and skills to meet the needs of his/her own cultural context can help to bridge the gap.

Trainers must be aware of cultural differences in terms of choice of language and examples (Antomarchi & Little 1993), and should encourage students to share information about cultural attitudes to the material heritage, as well as traditional approaches to preservation. Apart from considering the cultural context, it is important that both teachers and students recognize that the Western approach to conservation is not the only approach. Students should learn that there are long-established traditions elsewhere for the care of cultural material, such as in the care of monuments in Sri Lanka (Wijesuriya 2000) or the care of works of art on paper in Japan (eg Usami 1988, Zhong 1988). It is important that students learn about other traditions of preservation because not only may they have cultural meaning (see Labi 1993) but their application may affect the appearance, condition and significance of objects now housed in Western museums.

An excellent example of training that has responded to the cultural context is the PREMA programme which provides training in conservation management in sub-Saharan Africa. The programme was implemented by ICCROM in Rome in 1987 with the express aim of meeting local needs by training a core of African museum personnel who would eventually take over and run the programme in Africa. It has had considerable impact on the care of the cultural heritage in sub-Saharan Africa in that teaching has addressed local conditions and encouraged adaptation of established methods to local circumstances. Short courses have focused on reorganizing documentation and storage in a number of museums (Figs 51 and 52), seminars for museum directors have introduced the concepts of preventive conservation, and university-level training has generated a network of trained personnel throughout the area. Now, at the beginning of the twenty-first century there are two centres, one to serve Francophone countries and one for Anglophone countries, which will take the work forward (Godenou 1995, Bergeon 1996, Antomarchi & Little 1999).

Figure 51 PREMA: training for conservation management in Sub-Saharan Africa. Museum stores in Zimbabwe before reorganization (Courtesy of Gaël de Guichen).

Figure 52 Museum stores in Zimbabwe after reorganization by museum staff during a short training course organized by PREMA (Courtesy of Gaël de Guichen).

Developing conservation procedures

Conservators are concerned that much work undertaken by scientists does not have a direct and useful bearing on their ability to conserve objects. There are relatively few scientists working in the heritage field and some of them work on problems which are not directly relevant to the practice of conservation (Lasko & Lodewicks 1982) such as scientific research on objects, which is important in its own right because it may deepen understanding of significance, but it may not advance conservation.

> *To be specific, is, for example, the microscopical examination of paint cross-sections conservation science? My opinion is that only when conclusions are drawn from that examination which directly help the practice of conservation can it be truly called conservation science rather than the technical study of paint constituents and structure. (Tennent 1997, 15)*

A disappointingly small number of publications of scientific work on ancient and museum materials relate to conservation problems. De Guichen (1991) discovered that among his international colleagues the view was that only about 20% of scientists working in the heritage field actually produced research which was relevant to conservation practice, and after surveying conservation literature he concluded that almost half of the scientific articles were studies of structure and composition of objects, not of conservation. He went on to point out that while the conservator's reputation depends on his/her having successfully conserved objects, the scientist's rests on his/her publications and 'it is easier, quicker and less risky for a scientist to publish something on the composition of an object, its past history, rather than on methods of conservation which deal with the future of the object' (de Guichen 1991, 24).

Understanding of the composition of objects and the way they deteriorate is, of course, essential but this is only the first stage in developing an effective conservation process. Publications describing the composition and structure of materials provide important background but the scientific research needs to be taken much further, to develop conservation processes or materials and test them by application, if it is to be useful for conservation (Price 1996). The problem is illustrated by titles of papers such as 'Applications of qualitative trace element analyses: an interdisciplinary approach to materials conservation' (Sickels-Taves *et al* 1997) which is about analysis of the composition of objects containing organic materials; or 'Machine vision as a materials conservation technology' (Samajdar 1997) which is about using new techniques for examining condition; or 'Use of FT-IR in the conservation of twentieth-century objects' (Shearer & Doyal 1991) which is about the examination of early plastic objects. All of these papers discuss techniques for examining the nature and condition of the object, but these techniques do not, of themselves, bring about the conservation of the object. They are essential tools for diagnosis but do not prevent or cure deterioration any more than X-radiography cures broken limbs. Conservators cannot use this data to improve approaches to conservation of objects unless additional scientific work is undertaken.

For the development of an effective conservation treatment several further steps are needed, starting from establishing deterioration mechanisms and assessing the risks they pose to objects, moving to proposing a range of different ways in which the causes and effects of this deterioration might be minimized, then considering the risks of each approach to the types of object which might require conservation (risks to both material and conceptual integrity and to the potential for use), eliminating some approaches as being possibly damaging and finally testing the remaining ones. Gaining results from this form of research into conservation treatments is slow, and the latter stages frequently involve applied rather than pure science. The current position is that a comparatively large number of scientists undertake the investigation of ancient materials but few take the research further and evaluate the implications for conservation. This is partly because of the way scientific research is funded (with the main funding going to fundamental, not applied, scientific research) and the way in which scientists see the development of their careers, but it is also because there has been insufficient dialogue between scientists and conservators.

Organ recognized this problem when he discussed the kind of scientific support that conservators need when examining objects prior to choosing an appropriate conservation treatment. He described both the basic level of routine analysis, and the next level, 'the customized approach', where the conservator and the analyst together discuss the object and select the sample (Organ 1981). He then went on to discuss the top level, the full-scale study which involves several people, samples taken from different objects, the production of a publication and possibly a breakthrough in the form of development of a new technique, but he pointed out that this type of study is 'more likely to result in the creation of a new tool for use in authenticity studies' (ie for establishing material authenticity) (Organ 1981, 43). In other words, the application of existing analytical techniques is directly useful to the conservator in understanding the object and its conservation needs (or may be useful in understanding a material which could be used in a conservation treatment), but this is not the stuff of fundamental research. The fundamental research which attracts funding and results in publication is not undertaken with conservation in mind, and is often not directly applicable to conservation (though, of course, new analytical techniques may eventually be useful).

This confusion is compounded by the fact that, for some conservators, particularly archaeological conservators, technological examination has become almost more significant than conservation. Of course the information disclosed about an object during its examination and conservation can make an invaluable contribution to some forms of archaeological research (and this needs to be better recognized in the design of archaeological research projects) but if too much emphasis is put on the elucidation of ancient technology the process of conserving the object becomes devalued (Pye & Cronyn 1987). This in turn exacerbates the problem of conservation research: for some conservators (and possibly some funders too) technological research is seen as the only respectable thing to do and research into processes which lead to preservation has received scant attention.

That scientific analysis occupies a key or even dominant role in conservation (certainly in archaeological conservation) is one position, another is that science (materials science) is actually quite separate from conservation. This view is epitomized by the conference entitled *The Interface between Science and Conservation* held at the British Museum in 1997 (Bradley 1997a), at which some of the speakers certainly implied that 'science' and 'conservation' were two separate activities and that scientists worked independently to identify materials, or to provide solutions to conservation problems, after which the work on the objects themselves would be undertaken by conservators. There was the further implication that science and conservation might meet at 'the interface' but there was no real interaction between the two. However, unless the two are effectively interlinked, conservation processes will be developed empirically by conservators without access to scientific advice nor to the range of analytical equipment and expertise which working with scientists can offer.

A close working relationship is essential so that conservators can demonstrate the challenges and constraints of conservation which arise from both object and context, and from the need to use objects in interpretation and access. Dialogue is essential if scientists are to have a clear understanding of the multifaceted nature of objects, encompassing both material and conceptual characteristics. Scientific work is usually based on small samples taken from each object rather than on whole objects. Development of conservation treatments must be collaborative since working on samples does not reflect the material variability even within any one object. Further it is difficult (if not impossible) to simulate objects, thus there is a limit to the usefulness of testing on simulations; ultimately testing has to be undertaken by conservators on 'real' objects.

If research (and its associated funding) is directed largely into areas which do not result in improved conservation practice, then conservation will not progress as effectively and rapidly as it should. It is unfortunate that currently more emphasis is put on materials research than on conservation research. The exception is the topic of the museum environment where the fruits of research are leading to better preventive measures for the conservation of objects and collections.

What is needed to develop conservation processes?

The areas in which conservation research are needed are the development of theory and philosophy for archaeological conservation and objects conservation generally, the development of environmental management which is being effectively researched within the architecture and engineering fields, and the development of materials and methods for preventive and remedial conservation, which needs to be taken further through collaboration between conservators, conservation scientists and materials scientists.

One of the impediments to this kind of scientific research is the lack of funding (and status) for applied research. Materials and techniques exist, or are being developed in other fields, eg dentistry, mechanical engineering and polymer technology, which could be modified and applied to conservation of objects. External support in the form of research grants is needed for this kind of work. Support is needed also for long-term monitoring of the effects of preventive care regimes, or of particular preventive and remedial treatments, in order to complete research investigation, since testing on samples or simulations, even if artificially aged, will provide only indications of outcome, not reliable results.

Another field which has enormous potential for conservation and needs further research is information communication technology. This is becoming important for documentation but also has considerable potential for interpreting objects, for minimizing interventive approaches to repair and restoration through digital imaging of different possible options (and thus for aiding conservation decisions). Some research is already happening and this is undoubtedly a developing field.

Although many conservators would agree that there is not enough research leading to the improvement of thinking and practice, this lack has not been debated sufficiently and conservation will not progress until the situation is clarified. Developing the materials and procedures of conservation is just as important as developing the techniques of examination and elucidation. Conservation is an applied science and its procedures should be grounded in good scientific practice, and fully researched. It also requires a range of skills and expertise; thus conservation scientists can work no more effectively without the advice of conservators than conservators can work without the advice of scientists.

Current and future development of conservation

It does appear that a profession of conservation is now emerging, and once proposed accreditation schemes take effect they will do much to maintain good standards of practice (and to raise standards where necessary). The advancement of both discipline and profession is dependent on the quality of education and training in the field, and trainers are currently active in drawing on educational theory and discussing and developing effective education and training programmes. Together these will ensure good standards of practice. The advancement of the discipline is also clearly linked to development of scientific research, and it is important that the question of what types of research are relevant to conservation is fully debated and resolved if there is to be significant innovation of technique.

10 Communicating conservation

Introduction

Conservation will not be sustainable without public support (Foley 1999b). The aims of this chapter are to discuss the impact of cultural heritage conservation on the public (from politicians to school students), and whether conservation is understood and supported or whether more could be done by conservators to make the aims and constraints of conservation apparent, and to generate a feeling of *shared* concern for the heritage.

Obviously it is to the advantage of conservation and conservators that the aims and processes of conservation should be generally understood and supported by the public (here meaning individuals and groups who are not professionally involved in the care of cultural material). Most of conservation is supported directly or indirectly through the public's taxes but much of the damage is caused by the public, often entirely inadvertently. It is clear from the pattern of job reductions in museums that in-house conservation is frequently seen as less necessary than other types of in-house service, and it continues to be a largely invisible activity, partly because much of it is now undertaken by independent conservators. Better public and professional understanding and support would help to make conservation 'visible'.

Why communicate?

Many conservators feel that conservation of the cultural heritage is poorly understood and supported within the heritage professions and by the public. In fact little is known about what the public thinks about conservation or about attempts to raise public awareness. In 1989 Stone discussed this problem with reference to archaeology, pointing out that some had argued that 'archaeologists in Britain have little understanding (or even regard) for what the public thinks about the past or archaeology' and that 'experts deliberately hide archaeological information in high-flying academic jargon' (Stone 1989, 196). The crucial importance, and urgency, of conservation will not be absorbed by the public unless conservators take a more active role in engaging people with cultural heritage conservation. Conservators should have the confidence to show their enthusiasm for what they do and should be prepared to change their language, attitude and visibility in order to capture the public imagination.

The conservation of the natural heritage has aroused strong feelings among many members of the public, and has been effectively publicized. Should cultural heritage adopt some of the strategies used to promote the natural heritage? There is a possibility that being seen as over-scientific and dispassionate is losing support for the conservation of the cultural heritage. Is being scientific a deterrent, or is it possible that the science of preventive conservation, which is particularly well-developed, is less attractive than the science involved in investigating and learning about objects (perhaps similar to the exploration of behaviour of animals or genetic history in natural history specimens)? Should conservation involve the public far more in exploring objects and discovering the fascination for themselves?

Controversy

Controversy can be seen as damaging, but it is one way in which issues can be brought to the public attention. Conservation of cultural heritage should be prepared to use controversy rather than avoid it. Controversy shows that there are strong feelings about the cultural heritage.

Destruction of the natural and cultural heritage is normally condemned, and conservation is normally supported. The agents of destruction are, in general, considered to be clearly identifiable: poachers, hunters, loggers, looters, industrial-scale farming, felling of rain forest, war damage. But where nature conservation on the whole has a good press, cultural heritage conservation is often portrayed as being a barrier to access and to progress. Conservation agencies are too frequently accused of wanting to 'preserve in aspic' even though this is rarely the case. Perhaps even more negatively, *restoration* of cultural heritage objects is often portrayed as damaging and restorers are accused of being destroyers. Conservators need to explain the complexity of the problems they deal with, and be prepared to admit that they do not always get it right!

The controversies surrounding restoration of paintings and buildings which manifested themselves in the nineteenth century revive periodically when the conservation of a major work of art, or a work with religious or national significance is the focus of attention. Price reported on the outcry and the conflicting views aroused by the restoration, in 1970, of the medieval stone figures of the Virgin and Child, and proposals for the restoration of the Coronation of the Virgin on the West Front of Wells Cathedral in the UK (Price 2000a). Both the treatment of the Elgin Marbles (originally part of the Parthenon, Athens) in the British Museum in the 1930s (see St Clair 1999) and the restoration of Michelangelo's ceiling in the Sistine Chapel (in the Vatican, Rome) during the 1980s provide examples of widely publicized and international controversies, though it is interesting to note that the team working on the Sistine Chapel made considerable efforts to provide information about the work throughout the project (Mancinelli 1991, Mancinelli 1992). Many of these controversies arose from the meaning of the objects in terms of local, or national, cultural identity and conservators must realize that this produces emotions which colour reactions to conservation 'interference'.

There are numerous other instances, as indicated by book titles such as *The Ravished Image* (Walden 1985) or headlines such as 'Restorers are "ruining Titians"' (Jury 2000). Both Hedley (1986) and Bomford (1994a) discussed the nature of the controversy surrounding restoration of paintings, and the way in which restorers are alleged to cause damage, and Bomford commented 'Even today, there are conservators who can show a frightening lack of imagination or interest in the objects they are treating and analysing' and concluded that if the conservator is seen 'not as a defender but as a threat, then something has gone terribly wrong and we have a lot of explaining to do' (Bomford 1994a, 3, 4). It is dangerous if conservators are seen as having no interest in, or appreciation of, the objects they deal with. Communicating, not only enthusiasm for objects, but knowledge and understanding which can result from investigation and conservation, is vital.

In the archaeological context preservation can also be controversial, as in the case of the Bronze Age timber circle known as Seahenge, uncovered by the tides in 1999 at Holme-next-the-sea on the north coast of Norfolk, UK (see Chapter 7). Local people and interest groups such as druids had passionate feelings about the site and wanted the circle to be left *in situ*, but English Heritage decided, on the basis of specialist advice, that it must be removed for study and conservation before it was destroyed by erosion. This caused considerable anger and resentment among local people, and showed the need for English Heritage to take account of stakeholders' views, and to explain its actions more clearly. Experts from English Heritage came across as being over-scientific and dispassionate, and unsympathetic to the significance of the monument to the local population. 'If we want to involve communities in their own historic environment, we must accept their right to ask difficult questions and our obligation to engage in dialogue' (Miles 2000, 7). To which could be added the need to communicate the potentially fascinating information which may be yielded through studying the excavated timbers and central oak stump (Figs 53 and 54).

Figure 53 Seahenge: a contested site. A Bronze Age timber circle exposed between tides on the North Norfolk coast, UK. The circle was of great conceptual significance to some members of the public who wanted it left *in situ* (Courtesy of Francis Pryor).

Figure 54 Seahenge: potential loss of evidence. The remains of the peat in which the timbers were found were eroding rapidly. If the timbers had been left in position how long would they have survived? (Courtesy of N D Balaam).

A similar pattern of anger and frustration was involved when the highly damaged remains of Shakespeare's Rose Theatre were discovered near the Thames in London in 1989 (Wainwright 1989). Lack of effective communication between archaeologists and other stakeholders (actors and Shakespeare specialists) over the theatre's preservation led to a serious public controversy. This apparent lack of sympathy can be felt on a more personal level by house owners who are irked by the restrictions imposed by the conservation rules applied to historic or protected buildings, thus creating fertile ground for controversy and criticism. A reputation for being unsympathetic, restrictive or even damaging, makes it difficult to provide a clear message about the advantages of cultural heritage conservation. Those responsible for conservation need to be more aware of the public's concerns and to demonstrate the benefits of conservation in relation to the quality of people's lives (English Heritage 2000b).

Heritage conservation as a collective responsibility

Conservation of our cultural heritage should be a collective responsibility, and needs to be widely supported, but will not be without real (rather than token) involvement of the people who are interested in the heritage, however 'non-academic' their interests are. Politicians and decision-makers need to be convinced of this overall responsibility. Talley (1999) made the point that the highly successful Delta Plan (an exercise in dealing with the conservation backlog in museums in The Netherlands) owed much to the lobbying by conservators and the eventual willingness of politicians to take up the challenge.

Archaeology has gained enormously from the input of the public: members of the public have been responsible for discovering and reporting major finds, many have worked on excavations, and fruitful partnerships have been developed with metal detector users. Conservators should inform and involve as many people as possible in the business of conservation. In the museum everyone has a role to play in preventive conservation. Here, the aim is to create an overall environment which encourages preventive conservation throughout the museum, and involves all staff, as well as the visiting public and interested partners, such as the local community. The same kind of involvement could be generated on a wider scale.

The public is now familiar with the issues relating to nature conservation, and many people do apparently see this as a collective responsibility, but in general the public is much less informed about conservation of artefacts and buildings. Where artefacts are concerned this may be partly because conservators have been reluctant to popularize the subject for fear of providing 'recipe-book' conservation advice which might be used to the detriment of important artefacts or structures, but this perpetuates the view that only 'experts' can be involved, so undermining possible public interest and support.

There is also a common fear among conservators that people will be interested in conservation-restoration purely for monetary gain, in that antiques in good condition can be sold for more, and buildings with their original features preserved command a higher price. The support of a local community may be crucial in ensuring that a monument is protected, for example, explanation of the cultural values of local heritage may turn looters into protectors once a feeling of personal ownership has been created, as happened at the Moche tombs at Sipan, Peru (McManamon & Hatton 2000). Conservators should not ignore the fact that people are interested in the value of their personal heritage since a consequence of an awareness of the values of well cared for or maintained objects or structures ought to be a growing understanding of the value of conservation itself.

Accessibility of conservation

One of the key factors in successful communication about conservation is accessibility. Conservation is often seen as something done by 'experts' and is largely invisible behind the scenes, but it needs to be demystified and opened-up, not just as a collective responsibility but an activity in which everyone can share, whether it is caring for family heirlooms at home, or participating in clearing waterways and rebuilding walls in historic parks, or working professionally on objects in museum collections. Conservation should be communicated as a way of providing both intellectual and physical access to objects, by knowing more about them, and being able to handle and examine them. When designing ways to publicize conservation, such as exhibitions or publications, conservators must discard the jargon of the discipline and explain the aims and processes of conservation in a language that non-specialists can understand.

To communicate effectively with the public conservators need to understand that people are interested in the past, and thus in objects, in different ways. The impersonal local or national heritage (eg the local church, Stonehenge, Mount Vernon) may interest only some of the population (Merriman 1991). Appealing to people's involvement in their personal past, through family 'heirlooms' or their own collections, may be a more successful way of generating interest in conservation. Interest in personal collecting is responsible for the success of 'People's Shows' in museums, and 'clinics' during which curators or conservators provide information about, and advice on the care of, personal possessions.

Rewards of raised awareness

Publicizing conservation and raising public awareness are important since the rewards in the form of increased public support can only benefit conservation of the cultural heritage.

> *The benefits of such awareness efforts go far beyond the enlightenment of single individuals. Funding through federal, state and local agencies can be influenced by heightening the concern of both individuals and private-sector leaders. When the public gains a clearer understanding of conservation, both private and public collections are scrutinized through more informed eyes and the need for their protection takes on increased importance. And, of course, some are so inspired by what they learn that they decide to join the profession. (Podany & Maish 1993, 101)*

Communicating conservation may help the public to understand the complexities and constraints of conservation and restoration practice, and so minimize the negative impact of controversy. Publicizing conservation also means that conservators have to develop clear explanations for their actions, which may have the added benefit of making them re-evaluate elements of their practice.

Means of attracting public support

Using controversy

One of the techniques which the nature conservation lobby has used very effectively to attract the public's attention is to quantify the rate of loss of the tropical rain forest, or of individual species of plant, animal or bird. 'The house sparrow population has fallen by 65% since 1970, and by 90% in urban areas' (McCarthy 2000). 'Latest statistics show that we lost three out of every four skylarks during the last quarter of the twentieth century' (Donald 2000, 29). This message is usually followed by emphasizing how much the poorer we will all be if we lose the forest or the bird-song; for example 'the Government included the numbers of skylarks and other birds in a new "quality of life" index' (ibid).

When these tactics have been used by the supporters of cultural heritage conservation they have had considerable impact. The rescue lobby in the UK used them effectively in the 1970s in order to establish the need for rescue (salvage) archaeology, to deal with rapid destruction of archaeological sites. 'There is a similarly grim picture from Wiltshire. A survey made in 1964 of the 100 recorded Romano-British settlements showed that only 10 were still well-preserved' (Barker 1974, 29), and 'some 3,200 sites have been destroyed simply by the increase in forestry between 1960 and 1970' (Barker 1974, 32). In 1975 an influential and dramatic exhibition at the Victoria and Albert Museum in London, and a related publication, both entitled *The Destruction of the Country House*, drew attention to the large number of historic houses demolished between 1875 and 1975, particularly following the 1939–45 war (Strong *et al* 1975). Both these campaigns had considerable impact on professional and public opinion and led to investigation and preservation rather than destruction.

More recently the British Library chose to launch an Adopt-a-Book scheme to increase the funding for conservation, and announced that there were 300,000 books awaiting repair in order to make them usable again by readers, and that the total increased every year (Watson-Smyth 2000b). Since the public who use the Library will be directly affected if the books are not available because they are in poor condition, this same public may prove to be very supportive when appealed to in this way.

Perhaps conservation needs to make use of controversy by quantifying the loss of the cultural heritage, and challenging the public more often to think about what the quality of life would be like with no historic buildings or few ancient objects. This tactic worked well in The Netherlands where investigation and quantification of the backlog in documentation, preventive and remedial conservation in state museums produced the stark conclusion in 1990 that £125 million was needed to remedy the situation, and out of this grew the very impressive Delta Plan (Talley 1999). On the other hand, simply emphasizing loss may be a less successful message than emphasizing the gain in the quality of life if the richness of cultural heritage is conserved and accessible.

Newspapers

The press welcomes controversy, but unless the controversy focuses on the causes of damage, the impoverishment of the public's environment and loss of enjoyment caused by destruction, the message may not promote cultural heritage conservation. Unfortunately the press often promotes a negative message about conservation.

In 1988 the report on the management of the UK national collections (Bourn 1988, Sheldon 1988) gave rise to newspaper headlines about the Victoria and Albert Museum, London, such as 'V&A admits "disastrous failure"' and quotes such as 'Some 100,000 items in the V&A's prints and drawings collection likewise required urgent treatment. With present staffing levels it would take 200 years to complete that work' (Fletcher 1988). In this case the report certainly gave rise to controversy, but the image of the museums portrayed in the press was one of irresponsible neglect, thus making the museums the target of criticism rather than the real causes of damage such as pollution and lack of funds. There are other possible ways of looking at this. One is that management decisions have to be made to select some material for priority treatment (which means, of course, that some material does not get treated) but this is a sensible use of public funds, now that preventive conservation is increasingly effective. A second is that there may be an advantage in retaining an untreated 'control group' since remedial procedures are far from perfect.

Equally unfortunately, the fact that restoration is controversial means that it is restoration which is targeted by the media, and restorers' actions which are questioned.

> *Everywhere experts are questioned and derided, in politics and in every other profession including ours ... Newspapers ... are delighted to feature allegations of malpractice, but not at all interested in printing reasoned replies. (Bomford 1994a, 4)*

Conservators must accept that reasoned replies do not capture the public imagination, controversy must be met with equally strong feeling about the gains to be made through conservation. At present conservation is apparently understood no better, and examples which have adopted a minimal approach and have left the object or structure visually unchanged do not necessarily get good treatment by the press. The conservation of Wigmore Castle in

Herefordshire, UK (by English Heritage) was criticized by some journalists for costing £1 million to consolidate and make safe, eg 'Heritage £1M keeps castle in ruins' (Hamilton 1999). The intention was to leave it looking much as before, ruined, overgrown and magical (see Fig 18).

> *Wigmore will be maintained very much as it was, a site that still has a real sense of discovery, little changed from its natural state, though stable and accessible, perhaps understated but preserved without serious damage for posterity. (Coppack 1999, 69)*

It is probably safe to say that similar expenditure on restoration would have been just as controversial.

Looting and questions of repatriation are frequently reported but there is little or no discussion of broad conservation issues such as the loss of significance resulting from loss of context, or the gains from shared custodianship of repatriated objects. Conservators have to accept that looting is less important than the personal catastrophes which result from war and other disasters, but, nevertheless, they could do more to communicate the loss of identity which goes with loss of cultural heritage.

Television

Television is a powerful medium of communication, and the supporters of nature conservation have used it very effectively. Unfortunately the material heritage does not yet have the same emotive appeal as animals, and television programmes, if they feature conservation at all, have tended, just like newspapers, to focus on the conservation controversies such as those provoked by cleaning paintings or sculpture. In 1989 Stone discussed the role of television and radio in promoting archaeology and the fact that programmes then tended to be too academic and inaccessible. He urged that 'archaeology must be presented in a lively and exciting way' and 'if potentially large audiences are ignored ... archaeology will have missed a golden opportunity to influence the public' (Stone 1989, 200).

Much more adventurous coverage is now being given to archaeology, certainly on British television, and conservation should be able to build on this. A British television administrator speaking in 1999 said 'Even ten years ago it was considered impossible to show a programme about pre-twentieth century history. That has changed', and another said 'In the run up to the Millennium, people are asking more questions about who we are, and where we come from. It's fascinating to explain: all the bones and artefacts and bits of pottery have helped shape our destiny' (Sanai 1999). The new approach to televising archaeology is to make it both less obviously academic and more immediate.

One series of programmes which has become very popular on British television is Channel Four's *Time Team*. This is presented by a well-known television comedy actor, and involves the viewer in a range of three-day digs undertaken by professional archaeologists but displaying all the attendant dirt, frustration, dreadful jokes and occasional elation familiar to digging archaeologists.

> *Time Team has, in the words of Tim Taylor, the series producer, used 'quirkiness' to make it attractive. 'Before, the tradition in televising archaeology was to get some wonderful expert who would talk down to the audience. The key with us was to do it live, and risk failure'. (Sanai 1999)*

The success of this programme was acknowledged by Sir Neil Cossons during one of the interviews marking his appointment as Chairman of English Heritage in April 2000. 'The task now is to do for buildings what ... *Time Team* are doing for archaeology' (Binney 2000). Unfortunately, investigation and conservation of the excavated artefacts does not feature on this programme, though other aspects of the applications of scientific technique do. This appears to indicate either archaeologists' lack of appreciation of the potential (and the fascination) of conservation, or the failure of conservators to see the advantage of involvement and seize the opportunity, or perhaps both. It may also be because conservation can be slow to yield results (but the same could be said of excavation).

Conservators need to be ready to use the media by feeding newsworthy material to them. A very successful campaign directed at the media, 'Media Save Art', was organised by de Guichen through the International Centre for Conservation in Rome. The objective was to involve the media at all levels in communicating the importance of conservation, and the impact of the campaign was recognized by the International Institute for Conservation Keck Award in 1996 for promoting public understanding of conservation.

Publications

Several well-illustrated and attractive publications have been written by conservators for the general public, such as *Caring for Your Collections* (National Committee to Save America's Cultural Collections 1992), *Caring for Your Historic House* (Heritage Preservation and National Parks Service 1998) and *Looking after Antiques* (Plowden & Halahan 1987) but these are very likely to be sought after by people already interested in conservation issues because they are actively involved in building up collections, or in caring for family possessions. These publications perform a very useful role in providing information for those eager to know more, but the challenge is to reach people who are unaware of the problems of conserving the material heritage, or who are antipathetic because they do not understand the need for restrictions such as limitation of visitor numbers, or low light levels.

The Internet

The Internet offers great possibilities for publicizing conservation, and it is interesting to note that COOL (COnservation On Line) the website serving professional conservators, based at Stanford University, USA, introduced a section on *Conservation/Preservation Information for the General Public* in early 2000. It provided a list of sources which although 'intended for museum, library and archive professionals, in general ... do not require a technical background, and should be of use to the lay public' (COOL 2000). This, like the publications mentioned above, would attract people with some prior interest and who already wished to find out more about conservation, but there was no general statement about the aims and practice of conservation, simply a list of useful resources.

At the time of writing it was less easy to find information through an 'uninformed' search. A world-wide web search for 'conservation' undertaken in May 2000 yielded a large number of hits but the first ten were all about the natural heritage, and of the first thirty only two were clearly about the material heritage (number 18 was Conservation On Line - see above).

In the UK, the National Trust is responsible for the preservation of both landscape and historic buildings with their contents. A search of its website (National Trust 2000) yielded information about environment and conservation, and links to over forty bodies dealing with the natural environment, eleven dealing with the cultural heritage, and two bodies dealing with the conservation of the built heritage, but the three bodies which actually had 'conservation' in their titles dealt with environmental conservation. This masks the fact that the Trust has an effective and innovative material conservation department which is responsible for the enormous number and range of artefacts and buildings in its care. Other searches yielded broadly similar results. At the beginning of the twenty-first century cultural heritage conservation appears to be largely invisible on the Internet except to the already well-informed who recognize it when they see it, or who can refine a search appropriately.

Museums and the public

In 1999 the UK Museums Association adopted the following definition of a museum:

> *Museums enable people to explore collections for inspiration, learning and enjoyment. They are institutions that collect, safeguard and make accessible artefacts and specimens, which they hold in trust for society.* (Museums Journal *May 1999, 51)*

As in earlier (and other) definitions, preservation is a clearly stated function of museums. But in general the conservation activities within museums have been invisible, and have not been thought amenable to display or of interest to visitors: 'a museum function hitherto regarded as too specialist and inaccessible for general consumption' (*Museums Journal* December 1997, 33).

Increasingly during the last years of the twentieth century museums were making conservation of the cultural heritage visible by explaining conservation conditions in exhibition areas (eg low light levels), by holding open days, open evenings or open weeks, by advising visitors on the care of their own possessions and by organizing exhibitions on conservation itself. Communicating the nature of conservation to the public was deliberately encouraged by awards such as the International Institute for Conservation Keck Awards and the Conservation Awards made by the UK Museums and Galleries Commission. In 1999 the closing General Assembly plenary session of the ICOM Conservation Committee's meeting at Lyon, France, focused on 'Conservation and the Public.'

The way in which cultural heritage conservation was handled in museums contrasted with the approach to natural heritage conservation. Some natural history museums were active in adopting conservation as a potentially contentious issue, and some of their exhibitions caused objection and even outrage (Davis 2000). By contrast conservation of the cultural heritage has been explained and displayed, with few attempts to deal with the controversy of destruction (or of restoration).

> *As people continue to destroy the natural and built environment, do museologists have an obligation to adopt a more radical approach, in order to promote greater awareness and even positive action? (Davis 2000, 314)*

At the beginning of the twenty-first century there are signs that a more radical approach to cultural heritage conservation, even active promotion of conservation as a moral imperative, might develop from reactions to disasters such as earthquakes (the frescoes at Assisi), war damage (the library and museum at Sarajevo) or looting and illicit trade (Tubb 1995, Brodie *et al* 2000). Other factors involved in the destruction of the cultural heritage such as over-population, poverty, pollution, deep-ploughing and massive building works have not been properly addressed. Perhaps conservation has made comparatively little impact because conservators are, as yet, focused on showing how conservation is done rather than on alerting the public to the urgent need to deal with the ultimate sources of destruction. This is not to deny the importance of explaining how conservation is carried out. Although conservation has not, in the past, been considered worth displaying, visitors are usually fascinated by the processes, and transformations, involved in remedial conservation and restoration and this is an important way to capture interest and support.

Exhibitions and conservation

There are two obvious ways in which the concepts of conservation can be introduced to the public through museum exhibition: one is through explanatory material included in a mainstream exhibition (eg on the work of a particular artist; the material culture of a particular group), the other is through designing exhibitions focusing specifically on conservation.

Including information about conservation in an exhibition can be an important way to communicate some key issues to the public, but it is a pity that the conservation display board is often placed where it may be overlooked, for example right at the end of an exhibition. If part of the point of the explanation is, say, to let the public know why low light levels are needed for the safeguarding of these particular objects, the explanation should be placed near the beginning of the route round the display so that the visitors understand this before getting frustrated by what they may see as poor lighting. Equally, explanations about the investigation and treatment of individual objects have much more impact if they are included near the relevant object rather than in the form of a postscript just as the visitor leaves the exhibition. Including information on conservation issues needs to be seen as *adding* interest, rather than as imposing an irksome requirement on the curator and designer.

Podany and Maish (1993) described a very successful interactive exhibition organized at the J Paul Getty Museum which took the visitor from an explanation of conservation and its ethics, through a sequence of examination, treatment and environmental control. Members of the public produced a range of views about individual aspects of the exhibition but 'all expressed the desire to see more exhibits like this one on all facets of the conservation field' (Podany & Maish 1993, 108). Even if the only message which is retained by the visitor is the potential danger of touching and handling this is still beneficial. 'This display prompted several visitors to say that they will no longer be tempted to touch objects in a museum' (op cit 1993, 106).

Another example of a highly successful exhibition is the display at the Conservation Centre, at the National Museums and Galleries on Merseyside (Liverpool) UK, which won the European Museum of the Year Award in 1998 (Forrester 1997, Fraser 1997). This is possibly the first permanent exhibition designed to explain conservation to the public and is linked to a programme of studio tours, video sessions during which the public can watch and question a conservator at work, and activity sessions covering topics which are relevant to caring for family possessions (such as light and its effects, and mounting an embroidery). The exhibition was designed with great ingenuity to introduce the visitor to concepts such as what objects are made from, agents of decay and the detective work needed to understand the object, through different approaches to conservation treatment. The points are illustrated with eye-catching objects drawn from the collections, as diverse as paintings, bird-skins and a motor-bike, each with a brief case-history. The strength of this display is that the objects themselves are interesting, sometimes arresting, even though the focus is on conservation questions and solutions throughout. The approaches used by conservators have been broken down into clearly understandable concepts: 'doing as little as possible'; 'taking things apart, putting things together'; 'matching and patching'; 'to clean or not to clean'; and 'back to the drawing board'.

It is a weakness, though, that the limitations on space mean that much of the display on preventive conservation ('on the move', 'handled with care', 'prevention is better than cure' and 'the supporting act') is upstairs in a separate area, and here there are few objects. The visitor may not come away with an understanding that preventive activities form a crucial thread that runs through all conservation and that preventive procedures can be important alternatives to the more interventive laboratory or studio techniques. This is a pity since it is in the sphere of preventive conservation particularly that the public can make a contribution to the care of the heritage.

The success of both these ventures was that the teams had deconstructed conservation activities in order to draw out the essence to present to the public. 'An attempt to present any conservation treatment in a condensed, easily accessible, didactic display is risky. Many, perhaps most, of the concepts that become second nature to working conservators turn out to be quite difficult to distil for consumption by a general audience' (Podany & Maish 1993, 102). These exhibitions were also enhanced by the use of interactive exhibits, video and trained facilitators. The audience was apparently attracted and drawn in, partly through the immediacy of being able to press buttons and try things out for themselves, partly through a familiar medium, video, and partly through the stimulus and encouragement provided by trained facilitators.

Figure 55 Encouraging public interest: the queue of people waiting to see the Spitalfields Roman coffin at the Museum of London (Courtesy of Museum of London).

To transmit the message about conservation, should displays focus on the processes only? How important is the appeal of the objects themselves? A temporary exhibition, *Braving the Elements*, at the Pitt Rivers Museum in Oxford, UK, focused on 'fascinating and ingenious garments from around the world ... specially designed to beat the weather' and used them to convey a conservation message. 'Now these garments themselves are in need of protection if they are to be preserved for the future' and causes of deterioration and methods of conservation were explained (Pitt Rivers Museum 1997). An exhibition entitled *Creative Solutions* at the British Museum during 1999 aimed to demonstrate both the complexities of ethnographic materials and the methods used to conserve them. Research is needed into audience reactions before it is known whether emphasis on the fascination of objects dilutes the conservation message or strengthens it.

The fascination of the object was undoubtedly an important aspect of the success of the project at the Museum of London in 1999 which involved the public opening of a Roman lead coffin and the subsequent excavation of the skeleton and other contents (including textiles, bay leaves

Figure 56 The Spitalfields Roman coffin: conservators cleaning the coffin in the museum gallery (Courtesy of Museum of London).

and a glass vessel). 'For several days, with crowds of Museum visitors watching, conservators carefully recovered and recorded the contents of the coffin. It took a week to clear the coffin, a week in which 10,000 Londoners came to pay their respects to this unknown Roman' (Swain & Roberts 1999, 10) (Figs 55 and 56). How many of the visitors went away with a new interest in archaeology and conservation?

Archaeology and conservation

Just as museums have preservation as one of their core responsibilities, so archaeology is concerned with conservation. The Code of Conduct of the (UK) Institute of Field Archaeologists gives responsibility for conservation of the archaeological heritage as the second of its five main principles (IFA 1999), but there are many practising archaeologists who know comparatively little about the conservation of the artefacts they excavate or about the types of information that investigation of the artefacts can reveal, information that may be relevant to the interpretation of the object and of the activities which may have taken place on the site, and exciting to the public. Is neither side transmitting the interest and fascination to the other, and, equally, not capturing the public interest and support?

Just as it was thought that conservation was not a suitable subject for display in museums, so archaeologists do not necessarily think of harnessing the public's fascination with objects (and potentially with conservation) to their cause (thus, perhaps, the absence of conservation in *Time Team*). Archaeologists provide explanatory displays and leaflets at excavations, but conservation may not be featured. In an otherwise excellent leaflet for teachers about the English Heritage excavations at Whitby Abbey, conservation is simply mentioned with no explanation 'very fragile finds such as metal, leather or textiles need very special cleaning (without using water) and conservation to preserve them' (Walmsley & Buxton 1999, 3). In the same leaflet considerable coverage is given to soil sampling and dry and wet sieving. Admittedly conservation may not be visible on the site, but the finds are, and their potential to tell a story after examination and conservation by a conservator is fascinating. Merriman has commented that displays that deal with the archaeological process such as the highly successful Archaeological Resource Centre in York, UK, could be widened to 'show something of the normally hidden museum processes of collecting, documentation, conservation, storage and display, particularly by concentrating on the collecting instinct that is followed by such large numbers of the public' (Merriman 2000, 304).

Metal detecting began to emerge as an archaeological activity during the last quarter of the twentieth century. It was labelled 'treasure hunting' and its practitioners were shunned by archaeologists and conservators because of their association with potential, and actual, damage to archaeological contexts and objects. However, metal detector users are an important interest group, a group which, as Merriman argued (1991), may have previously participated in archaeological activities by acting as volunteers in excavations. Some archaeologists have developed links with detector groups, and the UK Treasure Act of 1997 and the Portable Antiquities Scheme (DCMS 1999) have provided an opportunity for dialogue. Although a few metal detector users may be unscrupulous, many are fascinated by, and knowledgeable about, the objects they discover. Conservators who refuse to show detector users how to clean their objects, adopting the attitude that 'a little knowledge is a dangerous thing' are failing to reach an audience which is eager for conservation advice, and who will probably care effectively for the objects as well as absorbing the wider message of conservation.

Cultural tourism

The popularity of cultural tourism has led to damage of the sites and monuments which tourists have come to see. Crowds in museums affect the relative humidity, crowds walking around monuments wear away the land-surface or the original paving or steps and the coaches they arrive in pollute the atmosphere and lead to further damage. Perier d'Ieteren (1998) has suggested that training tourist guides to discuss conservation issues would mean that restrictions on visitor numbers (or light levels) would be properly explained and the visitors would be more likely to support these conservation measures.

Nardi has been responsible for a number of conservation projects at tourist sites. Since 1984 his policy has been to encourage visitors to view the conservation or restoration work in progress, rather than to exclude them. 'The public is normally accustomed to associate the words "restoration" and "conservation" with rooms and monuments that are closed to outsiders, so making it impossible for them to visit things they'd like to see' (Nardi 1999, 44). If they are able to see the work the visitors go away with a positive view of conservation and a fascination for what they have seen. 'Public response has exceeded all expectations: long lines have formed under the scaffolding, guided tours have been requested for school groups and many people have returned periodically to follow the evolution of the work' (Nardi 1995, 10). This has had a very positive effect on the conservators too and they appreciate the contact with the public.

Looting of works or art and of antiquities, and consequent damage to sites, has flourished not only because of financial inducements for the looters but because the public, particularly those buying the objects, many of them tourists, have been ignorant of how they have been obtained and the damage the process has caused. Better communication with the public would mean that more politicians, more officials and more potential purchasers would be aware of the iniquities of the trade. Pressure has already forced some dealers to curtail some of their activities (though by no means yet enough) (Brodie *et al* 2000).

Joint custodianship

An important development involving museums, other heritage centres and indigenous communities is the collaboration which has developed, particularly in Australia and in North America. For the Native Americans 'heritage has become central to physical, cultural, economic and political survival' (Peers 1999, 24). Policies of forced assimilation during the twentieth century meant the loss of native languages, religious practices and traditional craft skills. In many cases the only information about how artefacts were made and used remained in museums in the form of the objects themselves and documentary evidence such as anthropologists' field notes. Museums became important sources of information and the focus of regeneration programmes (Simpson 1996). This has led to changes ranging from repatriation of some material through to consultation over the exhibition and conservation of the artefacts which remain in museums. 'Native people are asking ... to be treated as equal partners in caring for their cultural heritage' (Peers 1999, 27). This demands a new enabling role for curator and conservator (Bernstein 1992, Clavir 1996, Coote 1998, Peers 1999, Merriman 2000). McManamon and Hatton (2000) made the point that emerging from this development, of peoples repossessing their pasts, are new forms of professional practice embracing both traditional indigenous and Western practices, perspectives and technologies, and that each group of practitioners will learn from the other.

Involving the public

The public can be involved in conservation activities in a number of different ways. Volunteers play an important part in many aspects of museum work including conservation. Increasingly museums and libraries are inviting members of the public to 'adopt' objects in order to help pay for their care, such as the scheme launched by the British Library in 2000 (Watson-Smyth 2000b). There are activity holidays and courses which introduce the concepts of archaeology and conservation and provide participants with the opportunity to do 'hands-on' work. In the UK these include a range of excavations, and schemes such as National Trust volunteers schemes which involve participants in clearing and maintenance of garden and landscape features, and Cathedral Camps which involve student volunteers in helping to 'conserve these buildings and to preserve our heritage for others to enjoy' (Cathedral Camps 2000). Unemployed people and volunteers made an important contribution to the success of the Delta Plan in The Netherlands (Buisje 1995). A remarkable scheme in the USA entitled *Save Outdoor Sculpture!* sponsored by the National Museum of American Art (Smithsonian Institution) and Heritage Preservation, involves volunteers, including school students, in caring for public sculpture (SOS 1998). Other programmes in the USA aim to involve the public in protecting the archaeological heritage: 'as a profession we cannot work alone' (Lerner & Hoffman 2000, 245).

Capturing children's interest

Stone (1989) pointed out the importance of capturing children's interest and commitment. 'A school-based archaeological project is probably the most effective way of influencing a large section of the population with the limited funds available' (Stone 1989, 202). Introducing children to concepts of cultural heritage conservation will help to shape future opinion (Basile 1995) and to influence parents and teachers now, and many temporary conservation exhibitions have been linked to educational activities. The Conservation Centre (Merseyside) organizes a range of events and services linked to the permanent display including activities for schools and information packs for teachers. With aspects of archaeology as well as history included in the UK National Curriculum for schools, teachers needed more information about conservation. *In the Nick of Time* is an award-winning book written for teachers to help them to explain conservation, and to show children how conservation draws on art and science (which may encourage some children to consider it as a future career) (Newbery & Fecher 1994).

Performance art – a different way to involve the public

Public interest is often stimulated by presenting things in new ways, and one way of doing this is through performance art. Mark Dion has worked with found objects in different parts of the world. In 1999 he conducted *The Tate Thames Dig* in London, engaging volunteers in beachcombing on two sites on opposite sides of the river Thames, and in cleaning and organizing the finds (Beckwith 1999, Wilson 1999). The activities were on display on the lawn by the Tate Gallery, and engaged the public at many different levels. The volunteers were caught up in their experience of the work, members of the public offered comment and interpretation of the objects, or simply enjoyed the spectacle, professionals were drawn in to give lectures on objects, and on London and the Thames. Later in the year the finds were displayed, laid out in apparently classified groups, in a glass-fronted 'cabinet of curiosities' in the Tate Gallery itself, and visitors were encouraged to open drawers and examine the objects (McLaren 1999).

Was this art, was it archaeology, was it conservation? What it certainly did was to remove archaeological and conservation activities from the accepted conventions and stimulate new ways of seeing things. Rubber boots and white overalls became 'costume', volunteers acted the part of specialists, the fragments of discarded or lost objects found on the foreshore were transformed into art (Coles & Dion 1999). It was clear that some of the volunteers who might otherwise never have encountered archaeology or conservation had become very interested, and this performance was untrammelled by the usual academic or didactic expectations and allowed participants to discover the fascination of digging and of objects in their own way.

The trappings of conservation have engaged artists and designers in museums. At the Museum of London the design for the exhibition *London Bodies* was influenced by the plastic boxes and other packaging materials used in the Museum's conservation laboratories (Ganiaris pers. comm.). At Cheltenham Art Gallery installation artist Chris Dorsett created an exhibition featuring objects in boxes and costume in calico covers which drew attention to the 'silent' stored reserve collections (McKee 1999). When people from outside the museum world are excited by things largely ignored by those within the museum it can challenge accepted view-points. 'In a period of institutional self-reflection some curators value the artist as an agent of change' (Dorsett 1999, 14). Conservation may be given a higher profile both in and outside the museum.

Making links with natural heritage conservation

Conservation of the cultural heritage has the support of many people but has not captured the interest and commitment of as wide a range of the public as nature conservation. Is this because animals and plants have greater appeal and there is a genuine feeling of urgency in the need to preserve them (Carver 1996, de Guichen 1982), and because the advocacy of this cause has been more effective? Or is it because there is a conviction that flora and fauna are a non-renewable resource? Are they, in fact, any less renewable than archaeological remains? Before extinction is reached there is always the possibility of renewal in nature, whereas the destruction of a site or monument allows of renewal only through replication, or substitution (though some would argue that discovery of 'new' sites, and reassessment of value and significance of existing buildings and artefacts is, in effect, constantly adding to and renewing the cultural heritage (see Holtorf & Schadla-Hall 1999).

There are parallels between the conservation of the natural and cultural heritage. Many of the direct or indirect causes of damage are the same (pollution, changes in agricultural method, poverty and urbanization) and each exists together with deterioration and loss, so that study of the causes and impact of damage is an important characteristic of conservation. In each case it is now apparent that conserving the habitat, or context, is crucial to the meaningful survival of animal or plant species or of cultural artefact. Equally it is important to preserve diversity, not just single out the best-known and most obviously valued, but to preserve vernacular buildings, not just cathedrals, to preserve sparrows as well as eagles. In each case there is debate about the extent to which the preserved should also be used. Are buildings and industrial machinery more informative and better protected through being used rather than left idle or empty? Are animals better valued where controlled hunting is allowed rather than leaving them to roam destructively, become pests and be subjected to illicit slaughter?

There are, of course, many similarities between the conservation of natural heritage and cultural heritage museum collections. Both types of collection are stores of information, much of it as yet untapped. For example the potential for natural history collections to yield genetic information is just being realized and study of some material, eg birds' eggs, can yield information about environmental change (Duckworth *et al* 1993). More will be yielded by both types of collection as new questions are asked in the future. There is scope for shared research into conservation techniques since so many of the materials (some in their natural state, some converted to artefacts) are the same: skin, feathers, bone, plant material, rocks and minerals. In each case conservators must balance use with preservation. Brokerhof and Hummelen (1994) explored the similarities and differences in approaches to the conservation of art objects and natural history specimens, here too there is scope for collaboration between conservators.

Is it possible to distinguish, today, between the natural and the cultural heritage when in many cases their development has been interdependent? In archaeology, the study of past human environments (through examining botanical and animal remains) could be described just as accurately in terms of the natural heritage, ie the study of past landscapes or past flora and fauna. *The Manual of Heritage Management* (Harrison 1994) embraces both natural and cultural heritages in a single volume, and as long ago as 1972 UNESCO issued a single convention, *Concerning the Protection of World Cultural and Natural Heritage* (UNESCO 1972). This linking of cultural and natural heritage is clearly seen in conservation plans for historic sites which now require assessment of all aspects of significance including the landscape (or townscape) setting (Australia ICOMOS 1999). So, for example, the English Heritage project for the preservation of Stonehenge involves the restoration of the natural grassland of the plain by re-routing a modern road (English Heritage 2000a). In fact little of our natural heritage is unaffected by human agency and much of the 'wild' countryside is constructed and thus as much a part of the cultural heritage as a formal garden or a historic building.

In some cases the situation is more complex: the conservation of cultural and natural heritage are inextricably linked, but conflicting, for example where valued groups of old trees grow on ancient barrows or when colonies of bats, which are a protected species, inhabit buildings such as churches. Root action may be destroying the barrows, and bat urine and excrement may be damaging church wall paintings or memorials (Greeves 1989, Paine 1996). Clearly the various interest groups must adopt a wide view and work together to achieve a reasonable outcome for both natural and cultural heritage.

It is such a holistic approach that Greeves was arguing for when he urged that archaeological features should be saved as part of a landscape rather than removed through excavation. 'Archaeology has the potential to add a cultural dimension to the conservation debate, which has for too long been dominated by almost exclusive concern for the natural world' (Greeves 1989, 664). A wider approach to the environment was embraced in the aims of the English Heritage Strategy Review when it was announced in February 2000: 'The concept of the historic environment, as distinct from a fragmentary inheritance of monuments marked "of historic interest", is quite recent' and 'meshes with environmental and sustainability issues' (Morris 2000, 2). Taking this concept further, and referring to ecomuseums which make links between climate and landscape and the development of a local community, Davis (2000) suggested that a way forward for the promotion of cultural heritage conservation was not just to follow the example of natural history museums which have promoted natural heritage conservation, but to adopt a holistic approach which embraces human history, cultural heritage and the natural environment.

Linking cultural and natural heritage accords with policies on sustainability, now being adopted by museums, which draw together conservation in the widest sense through reducing energy consumption and conserving natural resources, while promoting the preventive care of cultural resources and using collections to explain the principles of conservation and sustainability (Cassar & Martin 1999).

Generating public interest

Capturing the interest of those who do not know anything about conservation, and do not visit museums or sites, is difficult through the Internet, books and leaflets, but an effective approach is to produce a publication which provides guidance for those who are involved in communicating conservation to the public. This was the idea behind *Ours for Keeps?*, a resource pack produced in 1997 by the UK Museums and Galleries Commission bringing together contributions and case studies from a wide range of conservators (MGC 1997). The pack provides an introductory guide to conservation, gives advice on different ways in which conservation can be included in a museum's agenda, how conservation can be explained through exhibition, education and other museum activities, and includes guidance on making presentations, dealing with the media and other aspects of publicity. The remaining sections of the pack provide advice on caring for a range of family treasures such as clocks, fans, photographs and silver. The concluding section gives details of useful organizations, suppliers and publications.

This is a versatile publication which can be used by different people in different ways. The style is informal and friendly and the emphasis is on making the advice as accessible as possible. It is particularly constructive to provide advice on how to present information: presentation matters in communicating ideas, but in fact there is very little simple guidance available. Not surprisingly, the pack has been very enthusiastically received, earning comments such as:

> *I enjoyed the idiosyncrasies. The individual character of the writer is discernible and this adds colour and a more personal approach to the advice and instructions given which make them more memorable ... I consider the document will make a significant impact and contribute to the profile of collection care. (Craven 1997)*

There are several possible ways in which cultural heritage conservation could attract the attention and support of the public, at least some of which would involve conservators in adopting a new attitude to public relations. Enthusiasm for the cultural heritage must be the starting point and the driving force: the public must *want* to save their heritage (professionals appealing to an uninterested public for support looks like self-interest).

Adopting a holistic approach and joining forces with the natural heritage conservation lobby could be an effective way in which to alert the public to the widespread effects of pollution and climate change, and to strengthen the lobby for preventive measures to ensure the sustainability of both natural and cultural heritage. Quantifying loss of, or damage to, cultural heritage through all types of disaster (including pollution, floods, fire, war and looting), and transmitting these statistics regularly, would also do more to emphasize the need for responsible protection and conservation of the cultural heritage. A compelling message about the damage to the general quality of all our lives is also needed: if lark-song can be considered important to our well-being, so can old buildings, or collections of easel paintings. The Internet particularly could be very effectively exploited to broadcast these messages and statistics and explain cultural conservation issues.

Adopting a radical approach and using controversy could allow the conservation-restoration lobby to transmit a starker message about the loss sustained through destructive factors including human ones. Instead of headlines saying 'Restorers destroy Titians' perhaps there should be more headlines like 'Car-users ravage much-loved monuments' or 'Thousands of family heirlooms lose value every year'. Since most people are interested in the monetary value of their own possessions conservators should be prepared to talk in terms of loss of this value through neglect, and retention or gain of value through responsible conservation (eg replacing wooden window frames with plastic ones not only affects authenticity, and may lower the value of a house, but may be a poor financial investment since plastics are known to deteriorate, especially when exposed to sunlight). Displaying the damage caused by inappropriate cleaning methods and materials can reinforce this message (this approach has convinced many metal detector users to stop cleaning metal work with vinegar or fruit sauces), and can encourage owners to adopt better methods.

In order to avoid 'compassion fatigue' it is important that a positive message is transmitted about the value of cultural heritage and the enrichment to be gained from saving and using it. More museums should consider displaying the fascination of the processes and effects of conservation, as the British Museum evidently intends to do in its new study centre due to open early in the twenty-first century. Displays could approach conservation in a number of different ways, either harnessing controversy, or showing the fascination of the objects themselves, or the excitement of revealing objects through conservation treatment.

Conservation in the future

Capturing the public imagination is obviously important for the future of conservation. The fact that some conservation and restoration activities are controversial is healthy because it indicates that there are strong feelings about the heritage. For some, possibly many, people the survival of the heritage is bound up with their quality of life, just as the existence of wild birds enriches lives. All those responsible for the cultural heritage, including conservators, need to consider different ways of drawing attention to the benefits of stewardship (and the costs of loss). Resources are never adequate for every worthy cause but if people are fired by the cultural heritage and genuinely involved in its care, conservation will be supported and promoted as an important element of sustainable development.

List of abbreviations

AIC	American Institute for Conservation of Historic and Artistic Works
ICCROM	International Centre for the Study of the Preservation and the Restoration of Cultural Property
ICOM	International Council for Museums
ICOM-CC	International Council for Museums – Conservation Committee
ICOMOS	International Council for Monuments and Sites
IIC	The International Institute for the Conservation of Historic and Artistic Works
MGC	Museums and Galleries Commission
NIC	National Institute for the Conservation of Cultural Property, Inc
UKIC	UK Institute for Conservation

References

Aben R H, 1995 Conservation of modern sculpture at the Stedelijk Museum, Amsterdam. In Heuman J (ed), *From Marble to Chocolate, The Conservation of Modern Sculpture,* Tate Gallery Conference September 1995, London, Archetype, 104–109.

Adam J-P, 1994 *Roman Building, Materials and Techniques* (trans Mathews A), London, Batsford.

AIC (American Institute for Conservation of Historic and Artistic Works) 1996 Code of Ethics and Guidelines for Practice. In *AIC Directory*, Washington DC, AIC, 24–31.

Aitchison K, 1999 *Profiling the Profession: A Survey of Archaeological Jobs in the UK*, London, English Heritage, Council for British Archaeology and Institute of Field Archaeologists.

Albano A, 1996 Art in transition. In Stanley-Price N, Talley M K and Vaccaro A M (eds), *Historical and Philosophical Issues in the Conservation of Cultural Heritage*, Los Angeles, The Getty Conservation Institute, 176–184.

Alexander E P, 1983 *Museum Masters, their Museums and their Influence*, Nashville, Tennessee, The American Association for State and Local History.

Allfrey M, 1999 Brodsworth Hall, the preservation of a country house. In Chitty G and Baker D (eds), *Managing Historic Sites and Buildings, Balancing Presentation and Preservation*, London, Routledge, 115–126.

Amoroso G G and Fassina V, 1983 *Stone Decay and Conservation*, Amsterdam, Elsevier.

Andrews G, 1991 *Management of Archaeological Projects*, London, English Heritage.

Antomarchi C and Little T L, 1993 Multicultural training: a rich mosaic or an impossible puzzle. In Bridgland J (ed), *Preprints for the 10th Triennial Meeting Washington, DC, USA 22-27 August 1993,* Paris, ICOM-CC, 711–714.

Antomarchi C and Little T L, 1999 2007: Exploring the impact of the PREMA training program on the conservation of cultural heritage. In Dardes K (ed), *Defining and Measuring Effectiveness in Education and Training, Proceedings of the interim meeting of the ICOM-CC Working Group on Training in Conservation and Restoration, Vantaa, Finland*, Vantaa, Evtek Institute of Art and Design, 11–19.

Appadurai A, 1986 Introduction: commodities and the politics of value. In Appadurai A (ed), *The Social Life of Things, Commodities in Cultural Perspective*, Cambridge, Cambridge University Press, 3–63.

Appelbaum B, 1987 Criteria for treatment: reversibility, *Journal of the American Institute for Conservation* 26, 65–73.

APT (Association for Preservation Technology) and AIC (American Institute for Conservation), 1991 *New Orleans Charter for Joint Preservation of Historic Structures and Artifacts.*

ARAAFU, 1996 *Les Aspects Esthétiques de la Restauration des Objects Archéologiques.* Paris, ARAAFU (Association des Restaurateurs d'Art et d'Archéologie de Formation Universitaire).

Arthur J, 1999 Cost benefit appraisal for the Wroxton Abbey state bed at Aston Hall, Birmingham Museums and Art Gallery. *CCF [Care of Collections Forum] Newsletter* September 1999, 2–4.

ArtWatch International, 1999 http://www.artwatchinternational.org

Ascenzi A, Bianco P, Fornaciari G and Rodriguez Martin C, 1980 Mummies from Italy, North Africa and the Canary Islands. In Cockburn A, Cockburn E and Reyman T A (eds), *Mummies, Disease and Ancient Cultures*, Cambridge, Cambridge University Press, 263–288.

Ashley-Smith J, 1995 *Consider the benefits, calculate the risks* http://palimpsest.stanford.edu/byauth/ashley-smith/benefits.html (accessed 27 January 2000).

Ashley-Smith J, 1999a *Risk Assessment for Object Conservation.* Oxford, Butterworth-Heinemann.

Ashley-Smith J, 1999b Reversibility – politics and economics. In Oddy A and Carroll S (eds), *Reversibility – does it exist?* British Museum Occasional Paper 135. London, British Museum, 129–134.

Ashurst J, 1990 Appendix 1: Limewashing. In Ashurst J and Dimes F G (eds), *Conservation of Decorative Stone*, Vol 2 London, Butterworth-Heinemann, 229–230.

van Asperen de Boer J R J, 1998 Some reflections upon the impact of scientific examination on art historical research. In Heumens E (ed), *Looking through Paintings, the Study of Painting Techniques and Materials in Support of Art-historical Research*, Baarn (Netherlands), Uitgeverij de Prom, and London, Archetype Publications, 13–18.

Atkins M J, Beattie J and Dockrell W B, 1993 *Assessment Issues in Higher Education*, London, Department of Employment.

Australia ICOMOS, 1999 *The Australia ICOMOS Charter for the Conservation of Places of Cultural Significance (The Burra Charter)* Adopted 1979, with revisions in 1981, 1988 and 1999, International Council for Monuments and Sites.

Avrami E, Mason R and de la Torre M (eds), 2000a *Values and Heritage Conservation*, Los Angeles, The Getty Conservation Institute.

Avrami E, Mason R and de la Torre M, 2000b Report on research. In Avrami E, Mason R and de la Torre M (eds), *Values and Heritage Conservation*, Los Angeles, The Getty Conservation Institute, 3–11.

Bacon A F and Brown A J E, 1996 A partnership in learning: the teaching institutions and the profession. In Cronyn J and Foley K (eds), *A Qualified Community: Towards Internationally Agreed Standards of Qualification for Conservation*, London, ICOM-CC, 39–44.

Baer N S, 1989 Introduction. In Baer N S (ed), *Training in Conservation*, New York, New York University Institute of Fine Arts, 3–6.

Baer N S and Snethlage R, (eds) 1997 *Saving Our Architectural Heritage, Conservation of Historic Stone Structures. Report on the Dahlem Workshop, Berlin 1996*, Chichester, John Wiley.

Baker D, 1999 Introduction: contexts for collaboration and conflict. In Chitty G and Baker D (eds), *Managing Historic Sites and Buildings: Reconciling Presentation and Preservation*, London, Routledge, 1–22.

Baker M T, 1995 Ancient Mexican rubber artifacts and modern American spacesuits: studies in crystallisation and oxidation. In Vandiver P B, Druzik J R, Madrid J L G, Freestone I C and Wheeler G S (eds), *Materials Issues in Art and Archaeology IV*, Pittsburgh, Pennsylvania, Materials Research Society, 223–232.

Baker M T and McManus E, 1992 History, care, and handling of America's spacesuits: problems in modern materials, *Journal of the American Institute for Conservation* 31, 77–85.

Baldwin A, 1999 The wayward paper object: artist's intent, technical analysis, and treatment of a 1966 Robert Rauschenberg diptych, *Journal of the American Institute for Conservation* 38, 411–428.

Balfour D, Metcalf S and Collard F, 1999 The first non-intrusive upholstery treatment at the Victoria & Albert Museum, *The Conservator* 23, 22–29.

Barclay M H, 1990 An art gallery conservator: roles and responsibilities. In Ramsay-Jolicoeur B A and Wainwright I N M (eds), *Shared Responsibility: Proceedings of a Seminar for Curators and Conservators, co-hosted by the National Gallery of Canada and the Canadian Conservation Institute, Department of Communications, October 1989*, Ottawa, National Gallery of Canada, 24–29.

Barclay R L (ed), 1997 *The Care of Historical Musical Instruments*, Edinburgh, Museums and Galleries Commission, Canadian Conservation Institute, ICOM Committee of Musical Instrument Museums and Collections.

Barclay R and Mathias C, 1989 An epoxy/microballoon mixture for gap filling in wooden objects, *Journal of the American Institute for Conservation* 28, 31–42.

Barker P, 1974 The scale of the problem. In Rahtz P A (ed), *Rescue Archaeology*, Harmondsworth, Penguin Books, 28–34.

Barnett R, 1994 *The Limits of Competence*, Buckingham, The Society for Research into Higher Education and the Open University Press.

Barrows H S and Tamblyn R, 1980 *Problem-based Learning: an Approach to Medical Education*, New York, Springer.

Bar-Yosef O, 1985 *A Cave in the Desert: Nahal Hemar 9000 year old finds*, Jerusalem, The Israel Museum.

Basile G ,1995 Les enfants peuvent aussi contribuer à la sauvegarde des œuvres d'art, *ICOM Committee for Conservation Study Series* 1, 11–12.

Bearat H, Fuchs M, Maggetti M and Paunier D (eds), 1997 *Roman Wall Paintings, Materials, Techniques, Analysis and Conservation*, Proceedings of the International Workshop, Fribourg, Institute of Mineralogy and Petrography.

Beckwith N, 1999 Tide marked, *Museums Journal* September 1999, 25.

Bell D, 1997 *Guide to International Conservation Charters*, Technical Advice Note 8, Edinburgh, Historic Scotland.

Bender B, 1998 *Stonehenge, Making Space* Oxford, Berg.

Benjamin W, 1936 (1992) The work of art in the age of mechanical reproduction. In Benjamin W, *Illuminations*, London, Fontana, 211–244. [Also http://pixels.filmtv. ucla. edu/gallery/web/julian_scaff/ benjamin/benjamin.html]

Berducou M, 1996 Introduction to archaeological conservation. In Stanley-Price N, Talley M K and Vaccaro A M (eds), *Historical and Philosophical Issues in the Conservation of Cultural Heritage*, Los Angeles, The Getty Conservation Institute, 248–259.

Bergeon S, 1996 Conservation/ restoration skills: is geographical transfer relevant? In Cronyn J and Foley K (eds), *A Qualified Community, Towards Internationally Agreed Standards of Qualification for Conservation*, London, ICOM-CC, 68–72.

Berger J, 1972 *Ways of Seeing*, London, BBC and Penguin Books.

Bernstein B, 1992 Collaborative strategies for the preservation of North American Indian material culture, *Journal of the American Institute for Conservation* 31, 23–29.

Bertholon R, 1996 Quelques réflexions sur la liberté des choix esthétiques dans la conservation-restauration des objects archéologiques. In *Les Aspects Esthétiques de la Restauration des Objects Archéologiques* Paris, ARAAFU (Association des Restaurateurs d'Art et d'Archéologie de Formation Universitaire), 49–51.

Besterman T, 1998 Saying what museums are for – and why it matters, *Museums Journal* April 1998, 37.

Binney M, 2000 Wake-up call from the history man, *The Times*, 3rd April 2000.

Blackwood B, 1970 *The Classification of Artifacts in the Pitt Rivers Museum, Oxford*, Oxford, University of Oxford, Pitt Rivers Museum.

Blades N, 1998 Gaseous pollution in the museum environment: the uses and abuses of monitoring campaigns, *The Conservator* 22, 44–48.

Blake M, 1999 Why are there not more women museum directors? In Moore K (ed), *Management in Museums*, London and New Brunswick, The Athlone Press, 185–207.

Blank S, 1990 An introduction to plastics and rubbers in collections, *Studies in Conservation* 35, 53–63.

Bomford D, 1994a Conservation and controversy, *IIC Bulletin* April 1994, 3–4.

Bomford D, 1994b Changing taste in the restoration of paintings. In Oddy A (ed), *Restoration, Is it Acceptable?*, British Museum Occasional Paper 99, London, British Museum, 33–40.

Bomford D, 1998 Introduction. In Hermens E (ed), *Looking through Paintings*, London, Archetype Publications, 9–12.

Born H, 1993 Multi-coloured antique bronze statues. In La Niece S and Craddock P (eds), *Metal Plating and Patination*, Oxford, Butterworth-Heinemann, 19–29.

Bourn J, 1988 *Management of the Collections of the English National Museums and Galleries. Report by the Comptroller and Auditor General*, London, Her Majesty's Stationery Office.

Bowman S (ed), 1991 *Science and the Past*, London, British Museum Press.

Boylan P J (ed), 1992 *Museums 2000: Politics, People, Professionals and Profit*, London and New York, Museums Association in conjunction with Routledge.

Boylan P J, 1993 Museum careers: ever changing and ever growing, *Museum International* 45(4), 4–7.

Bradley S, 1990 *A Guide to the Storage, Exhibition and Handling of Antiquities, Ethnographia and Pictorial Art*, British Museum Occasional Paper 66, London, British Museum Publications.

Bradley S (ed), 1997a *The Interface between Science and Conservation*, British Museum Occasional Paper 116, London, British Museum.

Bradley S, 1997b The impact of conservation science in the British Museum. In Bradley S (ed), *The Interface between Science and Conservation*, British Museum Occasional Paper 116, London, The British Museum, 1–8.

Brajer I, 1999 Aspects of reversibility in transferred wall paintings. In Oddy A and Carroll S (eds), *Reversibility - does it exist?* British Museum Occasional Paper 135, London, British Museum, 63–72.

Bray W, 1993 Techniques of gilding and surface-enrichment in pre-Hispanic American metallurgy. In La Niece S and Craddock P (eds), *Metal Plating and Patination: Cultural, Technical and Historical Developments*, Oxford, Butterworth-Heinemann.

Brenni P, 1999 Restoration or repair? The dilemma of ancient scientific instruments. In Oddy A and Carroll S (eds), *Reversibility – does it exist?* British Museum Occasional Paper 135, London, British Museum, 19–25.

Brent M, 1996 The rape of Mali. In Vitelli K D (ed), *Archaeological Ethics*, Walnut Creek CA, Altamira Press, 163–175.

Brill R H, 1978 The use of equilibrated silica gel for the protection of glass with incipient crizzling, *Journal of Glass Studies* 20, 100–118.

Brimblecombe P, 1994 The balance of environmental factors attacking artifacts. In Krumbein W E, Brimblecombe P, Cosgrove D E and Staniforth S (eds), *Durability and Change. The Science, Responsibility, and Cost of Sustaining Cultural Heritage. Report of the Dahlem Workshop, December 1992*, Chichester, John Wiley, 67–80.

Brinch Madsen H, 1987 Artifact conservation in Denmark at the beginning of the last century. In Black J (ed), *Recent Advances in the Conservation and Analysis of Artifacts*, London, Summer Schools Press, 343–345.

Bristow I, 1990 An introduction to the restoration, conservation and repair of stone. In Ashurst J and Dimes F G (eds), *Conservation of Decorative Stone*, Vol 2, London, Butterworth-Heinemann, 1–18.

Brockbank A and McGill I, 1998 *Facilitating Reflective Learning in Higher Education,* Buckingham, Society for Research into Higher Education and the Open University Press.

Brodie F, 1994 Clocks and watches, a reappraisal? In Oddy A (ed), *Restoration, Is it acceptable?* British Museum Occasional Paper 99, London, British Museum, 27–32.

Brodie N, Doole J and Watson P, 2000 *Stealing History: The Illicit Trade in Cultural Material*, Cambridge, The McDonald Institute for Archaeological Research.

Brokerhof A W and Hummelen I, 1994 Is there a line and where do we draw it? Restoration ethics of natural history collections. In Verschoor H and Mosk J (eds), *Contributions of the Central Research Laboratory to the Field of Conservation and Restoration*, Amsterdam, Central Research Laboratory for Objects of Art, 43–42.

Brommelle N S, 1956 Material for a history of conservation, *Studies in Conservation* 2, 176–188.

Brommelle N S, 1964 The Russell and Abney Report on the action of light on water colours, *Studies in Conservation* 9, 140–151.

Brommelle N S and Pye E M, 1984 Introduction. In Brommelle N S, Pye E M and Thomson G (eds), *Adhesives and Consolidants. Preprints of the Contributions to the Paris Congress, 2-8 September 1984*, London, IIC, 3–4.

Brooks M, Clark C, Eastop D and Petcheck C, 1994 Restoration or conservation – issues for textile conservators, a textile conservation perspective. In Oddy A (ed), *Restoration, Is it Acceptable?* British Museum Occasional Paper 99, London, British Museum, 103–122.

Brooks M and Fairbrass S, 1997 Literacy in science: using the language in conservation. In Bradley S (ed), *The Interface between Science and Conservation*, British Museum Occasional Paper 116, London, British Museum, 31–38.

Brooks M, Lister A, Eastop D and Bennett T, 1996 Artifact or information? Articulating the conflicts in conserving archaeological textiles. In Roy A and Smith P (eds), *Archaeological Conservation and its Consequences. Preprints of the Contributions to the Copenhagen Congress*, London, IIC, 16–21.

Brorson Christensen B, 1970 *The Conservation of Waterlogged Wood in the National Museum of Denmark*, Copenhagen, The National Museum of Antiquities.

Bruce-Mitford R, 1978 *The Sutton Hoo Ship Burial, Vol 2 Arms, Armour and Regalia*, London, British Museum Publications.

Bruce-Mitford R and Evans A (eds), 1983 *The Sutton Hoo Ship Burial Vol 3 Part 2*, London, British Museum Publications.

Buck R D, 1971 What is condition in a work of art? *Bulletin of the American Group of the International Institute for Conservation* 12, 63–67.

Budd M, 1995 *Values of Art: Pictures, Poetry and Music*, London, Penguin Books.

Buisje J A, 1995 Aspects to the large-scale operation to save the Dutch cultural heritage activities at the Rijksmuseum voor Volkenkunde, *ICOM Committee for Conservation Study Series* 1, 6–8.

Burke J W (ed), 1989 *Competency-Based Education and Training*, London, New York, Philadelphia, The Falmer Press.

Butterfield F and Eaton L, (eds) 1991 *Paper and Textiles, the Common Ground. Preprints of the Scottish Society for Conservation and Restoration Conference*, Edinburgh, SSCR.

Cackett S, 1992 Disaster planning. In Thompson J M A (ed), *The Manual of Curatorship*, Oxford, Butterworth-Heinemann, 487–490.

Calnan C, 1992 The conservation of Spanish gilt leather. In Hodges H, Mills J and Smith P (eds), *Conservation of the Iberian and Latin American Cultural Heritage. Preprints of the Contributions to the Madrid Congress*, London, IIC, 23–26.

Campbell L, 1998 The conservation of Netherlandish paintings in the fifteenth and sixteenth centuries. In Sitwell C and Staniforth S (eds), *Studies in the History of Painting Restoration*, London, Archetype Publications in association with the National Trust, 15–26.

Cannon-Brookes P, 1988 The role of the scholar-curator in conservation, *Museum Management and Curatorship* 7, 323–325.

Carlyle L, 1998 Design, technique and execution: the dichotomy between theory and craft in nineteenth-century British instruction manuals on oil painting. In Hermens E (ed), *Looking through Paintings, the Study of Painting Techniques and Materials in Support of Art-historical Research*, Baarn (Netherlands), Uitgeverij de Prom, and London, Archetype Publications, 19–28.

Carver M, 1996 On archaeological value, *Antiquity* 70, 45–56.

Casey P J and Cronyn J M, (eds) 1980 *Numismatics and Conservation*. Occasional Paper 1, Durham, University of Durham Department of Archaeology.

Cassar M, 1995 *Environmental Management*, London and New York, Routledge.

Cassar M, 1998 *Cost/benefits Appraisals for Collections Care*, London, Museums and Galleries Commission.

Cassar M and Martin D, 1999 Update on environment: The route to sustainability, *Museum Practice* issue 12, Vol 4(3), 58–85.

Cathedral Camps, 2000 *Cathedral Camps, Summer 2000* (publicity leaflet; information also at www.cathedralcamps.org.uk).

Caygill M L, 1993 The protection of national treasures at the British Museum during the First and Second World Wars, *Material Issues in Art and Archaeology* III, Pittsburgh, Materials Research Society, 29–40.

Cennino d'Andrea Cennini, *Il Libro d'Arte (The Craftsman's Handbook)* trans 1933 Thompson D V, (1960) New York, Dover Publications.

Clark I C and Weaver M E, 1982 Conservation, decision-making and management, *Museum* 34(1), 21–30.

Clark K, (ed) 1999 *Conservation Plans in Action, Proceedings of the Oxford Conference*, London, English Heritage.

Clavir M, 1994 Preserving conceptual integrity: ethics and theory in preventive conservation. In Roy A and Smith P (eds), *Preventive Conservation, Practice Theory and Research. Preprints of the Contributions to the Ottawa Congress September 1994*, London, IIC, 149–153.

Clavir M, 1996 Reflections on changes in museums and the conservation of collections from indigenous peoples, *Journal of the American Institute for Conservation* 35, 99–107.

Clavir M, 1999 Museum changes to First Nations objects, and their physical and conceptual reversibility. In Oddy A and Carroll S (eds), *Reversibility – does it exist?* British Museum Occasional Paper 135, London, British Museum, 169–172.

Clavir M, Johnson E and Shane A, 1986 A discussion on the use of museum artifacts by their original owners. In Barclay R, Gilberg M, McCawley J and Stone T (eds), *Symposium 86, the Care and Preservation of Ethnological Materials*, Ottawa, CCI, 80–89.

Cockburn A, 1980 Introduction. In Cockburn A, Cockburn E and Reyman T A (eds), *Mummies, Disease and Ancient Cultures*, Cambridge, Cambridge University Press, 1–14.

Coles A and Dion M, (eds) 1999 *Archaeology*, London, Black Dog Publishing.

Coles J M, 1979 *Experimental Archaeology*, London, Academic Press.

Coles J M, 1987 The preservation of archaeological sites by environmental intervention. In Hodges H W M (ed), *In Situ Archaeological Conservation*, Mexico, The Getty Conservation Institute/INAH, 32–55.

Collier J, 2000 *Pharmacopoeia, A Science and Art Collaboration by Dr Liz Lee and Susie Freeman*, Notes to accompany an exhibition at Contemporary Applied Arts, London, February 2000.

Conservation Forum, 1998 *Guidelines for the Commissioning and Undertaking of Conservation Work*, London, Conservation Forum.

Cook C and Grattan D, 1990 A method for calculating the concentration of PEG for freeze-drying waterlogged wood. In Hoffmann P (ed), *Proceedings of the 4th ICOM-CC Working Group on Wet Organic Archaeological Materials Conference*, Bremerhaven, ICOM, 239–252.

Cooke B, 1990 Fibre damage in archaeological textiles. In O'Connor S and Brooks M (eds), *Archaeological Textiles*, UKIC Occasional Papers 10, London, UKIC, 5–14.

Cooke W D, Babakhani A and Hillyer L, 1996 The cleaning of degraded linen, *The Conservator* 20, 3–14.

COOL, 2000 Conservation/ Preservation Information for the General Public, *Conservation on Line* http://palimpsest.stanford.edu/bytopic/genpub (accessed 25th May 2000).

Cooper M, 1998 *Laser Cleaning in Conservation, An Introduction*, Oxford, Butterworth-Heinemann.

Coote K, (ed) 1998 *Care of Collections*, Committee for Aboriginal and Torres Straits Islander Keeping Places and Cultural Centres, Sydney, Australian Museum.

Coppack G C, 1999 Setting and structure: the conservation of Wigmore Castle. In Chitty G and Baker D (eds), *Managing Historic Sites and Buildings, Reconciling Presentation and Preservation*, London, Routledge, 61–70.

Coremans P, 1969 The training of restorers. In ICOM (International Council for Museums) *Problems of Conservation in Museums.* A selection of papers presented to the joint meeting of the ICOM Committee for Museum Laboratories and the ICOM Committee for the Care of Paintings, held in Washington and New York, September 1965. Paris, Editions Eyrolles, and London, George Allen and Unwin, 7–32.

Corfield M, 1992 Conservation documentation. In Thomson J M A (ed), *The Manual of Curatorship*, Second edition, London, Butterworth-Heinemann, 229–233.

Corfield M, Keene S and Hackney S, (eds) 1989 *The Survey: Conservation Facilities in Museums and Galleries*, London, UKIC.

Cornforth J, 1986 Introduction. In Sandwith H and Stainton S (compilers), *The National Trust Manual of Housekeeping*, London, Penguin Books in Association with the National Trust, 13–15.

Cossons N, 1992 Rambling reflections of a museum man. In Boylan P J (ed), *Museums 2000, Politics, People, Professionals and Profit*, London and New York, Museums Association in conjunction with Routledge, 123–147.

Cossons N, 1994 Scholarship or self-indulgence? In Kavanagh G (ed), *Museum Provision and Professionalism*, London and New York, Routledge, 237–246.

Craddock P and Bowman S, 1991 Spotting the fakes. In Bowman S (ed), *Science and the Past*, London, British Museum Press, 141–157.

Craddock P and Giumlia-Mair A, 1993 Beauty is skin-deep, evidence for the original appearance of classical statuary. In La Niece S and Craddock P (eds), *Metal Plating and Patination*, Oxford, Butterworth-Heinemann, 30–38.

Cramer F, 1994 Durability and change: a biochemist's view. In Krumbein W E, Brimblecombe P, Cosgrove D E and Staniforth S (eds), *Durability and Change. The Science, Responsibility, and Cost of Sustaining Cultural Heritage. Report of the Dahlem Workshop, December 1992*, Chichester, John Wiley, 19–26.

Craven T, 1997 quoted in the publicity leaflet for *Ours for Keeps? A resource pack for raising awareness of conservation and collection care*, London, Museums and Galleries Commission.

Creigh-Tyte S, Dawe G and Stocks S, 2000 *'The White Book'. Option Appraisal of Expenditure Decisions, A Guide for the Department of Culture, Media and Sport and its Sponsored Bodies*, Technical Paper 2, London, DCMS Finance Division.

Cronyn J M, 1990 *The Elements of Archaeological Conservation*, London, Routledge.

Cronyn J and Foley K, 1996 Standards of qualifications in conservation: is commonality either necessary or possible? In Cronyn J and Foley K (eds), *A Qualified Community, Towards Internationally Agreed Standards of Qualification for Conservation. Proceedings of the Interim Meeting of the Working Group on Training in Conservation and Restoration of the ICOM Committee for Conservation Maastricht, April 1995*, London, ICOM-CC Working Group on Training in Conservation and Restoration, 5–10.

Cronyn J and Pye E, 1988 The training of conservators: unity in diversity. In Todd V (ed), *Conservation Today. Papers presented at the UKIC 30th Anniversary Conference 1988*, London, UKIC, 21–23.

Crowther D, 1989 Archaeology, material culture and museums. In Pearce S (ed), *Museum Studies in Material Culture*, London and New York, Leicester University Press, 35–46.

Cruickshank P, Pullan M and Potter J, 1999 Recent treatments of painted Egyptian shrouds: the influence of condition and intended role, *The Conservator* 23, 37–44.

Cruickshank P and Tinker Z (eds), 1995 *Starch and Other Carbohydrate Adhesives for Use in Textile Conservation*, London, UKIC.

Curnoe G, 1990 Owning ideas, appropriation and protective collectives. In Ramsay-Jolicoeur B A and Wainwright I N M (eds), *Shared Responsibility, Proceedings of a Seminar for Curators and Conservators, National Gallery of Canada, 26th-28th October 1998*, Ottawa, National Gallery of Canada, 136–140.

Curtin B R, 1990 Is conservation ready for artificial intelligence? *The Abbey Newsletter* 14, 1–2.

Daintith C, 1995 A consolidation treatment for ethnographic pottery from New Guinea. In Hill L and Giles S (eds). *Where to Start, Where to Stop, Papers from the British Museum / MEG Ethnographic Conservation Colloquium*, Museum Ethnographers Group Occasional Papers 4, Hull, Museum Ethnographers Group, 121–130.

Daly Hartin D, 1990 An historical introduction to conservation. In Ramsay-Jolicoeur B A and Wainwrigh, I N M (eds). *Shared Responsibility. Proceedings of a Seminar for Curators and Conservators, National Gallery of Canada, 26th-28th October 1998*, Ottawa, National Gallery of Canada, 30–38.

Daniel G, 1975 *A Hundred and Fifty Years of Archaeology*, London, Duckworth.

Daniels V, 1989 Trends in the application of analysis in the museum laboratory. In Baer N S, Sabbioni C and Sors A I (eds), *Science, Technology and European Cultural Heritage. Proceedings of the European Symposium, Bologna, Italy, 13–16 June 1989*, London, Butterworth-Heinemann, 269–277.

Daniels V, 1999 Imperfect reversibility in paper conservation. In Oddy A and Carroll S (eds), *Reversibility – does it exist?* British Museum Occasional Paper 135, London, British Museum, 47–52.

Daniels V and Lohneis G, 1997 Deterioration of sugar artifacts, *Studies in Conservation* 42(1), 17–26.

Darrah J A, 1987 Metal threads and filaments. In Black J W (ed), *Recent Advances in the Conservation and Analysis of Artifacts*, Institute of Archaeology, Jubilee Conservation Conference Papers, London, Summer Schools Press, 211–222.

Davis P, 2000 Museums and the promotion of environmental understanding and heritage conservation. In McManamon F P and Hatton A (eds), *Cultural Resource Management in Contemporary Society, Perspectives on managing and presenting the past*, London and New York, Routledge, 310–318.

Davison S, 1984 A review of adhesives and consolidants used on glass antiquities. In Brommelle N S, Pye E M, Smith P and Thomson G (eds), *Adhesives and Consolidants. Preprints of the Contributions to the Paris Congress September 1984*, London, International Institute for Conservation, 191–194.

DCMS (Department of Culture Media and Sport), 1999 *Portable Antiquities Annual Report 1997-98*, London, DCMS.

DCMS (Department of Culture Media and Sport), 2000 *Guidance: Setting out the Composition, Definitions and Interpretation of the PI's.* Report prepared by Deloitte & Touche and LORD Cultural Resources and Planning.

Del Monte M and Sabbioni C, 1987 A study of the patina called *scialbatura* on imperial Roman marbles, *Studies in Conservation* 32, 114–121.

DiMaggio P, 1994 The American art director as professional: results of a survey. In Kavanagh G (ed), *Museum Provision and Professionalism*, London and New York, Routledge, 154–163.

Dollery D, 1994 A methodology of preventive conservation for a large expanding and mixed archaeological collection. In Roy A and Smith P (eds), *Preventive Conservation, Practice Theory and Research. Preprints of the Contributions to the Ottawa Congress September 1994*, London, IIC, 69–72.

Dollery D and Henderson J, 1996 Conservation records for the archaeologists? In Roy A and Smith P (eds), *Archaeological Conservation and its Consequences. Preprints of the Contributions to the Copenhagen Congress*, London, IIC, 43–47.

Donald P, 2000 The case of the missing skylarks, *Birds: The magazine of the Royal Society for the Protection of Birds* Summer 2000, 29–30.

Donnithorne A, 1988 Early approaches to the conservation of works of art on paper, cleaning, repair and restoration. In Daniels V (ed), *Early Advances in Conservation*, British Museum Occasional Paper 65, London, British Museum, 15–26.

Doré J, 1978 The conservation of two eighteenth-century court mantuas, *Studies in Conservation* 23, 1–14.

Dorsett C, 1999 Absence makes the heart grow fonder, *Museums Journal* January 2000, 14.

von Droste B and Bertilsson U, 1995 Authenticity and world heritage. In Larsen K E (ed), *Nara Conference on Authenticity in Relation to the World Heritage Convention, November 1994*, Tokyo, UNESCO World Heritage Centre; Agency for Cultural Affairs (Japan); ICCROM; ICOMOS, 3–16.

Drumheller A and Kaminitz M, 1994 Traditional care and conservation, the merging of two disciplines at the national museum of the American Indian. In Roy A and Smith P (eds), *Preventive Conservation, Practice Theory and Research. Preprints of the Contributions to the Ottawa Congress September 1994*, London, IIC, 58–60.

Drury M, 1988 The restoration at Calke Abbey, *Journal of the Royal Society of Arts* 136, 490–499.

Drysdale L, 1988 The eternal triangle: relationship between the object, the conservator and the client. In Todd V (ed), *Conservation Today. Papers presented at the UKIC 30th Anniversary Conference 1988*, London, UKIC, 19–20.

Drysdale L, 1999 The language of conservation: applying critical linguistic analysis to three conservation papers. In Bridgland J (ed), *ICOM Committee for Conservation Preprints of the 12th Triennial Meeting Lyon 29 August – 3 September 1999*, London, James & James, 161–165.

Drysdale L and Rowe A, 1998 The primacy of context, conservation of decorative arts from English Heritage sites. In Wright M and Player-Dahnsjö Y M T (eds), *Site Effects, The Impact of Location on Conservation Treatments*, Edinburgh, The Scottish Society for Conservation and Restoration, 77–82.

Duckworth W D, Genoways H H and Rose C L, 1993 *Preserving Natural Science Collections: Chronicle of Our Environmental Heritage*, Washington DC, National Institute for the Conservation of Cultural Property, Inc.

Durham A (ed), 1995 *Lining and Backing, The Support of Paintings, Paper and Textiles*, London, UKIC.

Durnan N, 1997 An assessment of the 'Lime Method' of conservation of exterior limestone sculpture. In Lindley P (ed), *Sculpture Conservation, Preservation or Interference?*, Guildford, Scolar Press, 83–94.

Dykstra S W, 1996 The artist's intentions and the intentional fallacy in fine arts conservation, *Journal of the American Institute for Conservation* 35, 187–218.

Eastop D, 1998 Decision-making in conservation: determining the role of artifacts. In Timar-Balazsy A and Eastop D (eds), *International Perspectives on Textile Conservation*, London, Archetype, 43–46.

Eastop D and Brooks M, 1996 To clean or not to clean: the value of soils and creases. In *ICOM Committee for Conservation. Preprints of the 11th Triennial Meeting, Edinburgh, September 1996*, London, James & James, 687–691.

Edson G and Dean D, 1994 *The Handbook of Museums* London and New York, Routledge.

Ellam D, 1999 More than gleaming machines: perspectives on the conservation of industrial collections, *IIC Bulletin* April 1999, 2–4.

Elston M 1998 A corrective treatment on a 6th century BC Attic Krater. In Paterakis A (ed), *Glass, Ceramics and Related Materials*, Interim Meeting of the ICOM-CC Working Group Vantaa, Finland, EVTEK Institute of Arts and Design, 106–114.

Emery I, 1966 *The Primary Structure of Fabrics: An Illustrated Classification*, Washington DC, Textile Museum.

Encyclopaedia Britannica, 1910–11, eleventh edition, New York, The Encyclopaedia Britannica Company.

English Heritage, 2000a *Stonehenge World Heritage Site Management Plan*, London, English Heritage.

English Heritage, 2000b *Discussion Paper 5, Enriching. Review of Policies Relating to the Historic Environment*, London, English Heritage.

Eraut M, 1992 Developing the knowledge base: a process perspective on professional education. In Barnett R (ed), *Learning to Effect*, Buckingham, Society for Research into Higher Education and Open University Press.

Erhardt D and Mecklenburg M, 1994 Relative humidity re-examined. In Roy A and Smith P (eds). *Preventive Conservation, Practice, Theory and Research. Preprints of the Contributions to the Ottawa Congress September 1994* London, IIC, 32–38.

Erhardt D, Mecklenburgh M F, Tumosa C S and McCormick-Goodhart M, 1997 The determination of appropriate museum environments. In Bradley S (ed), *The Interface Between Science and Conservation*, British Museum Occasional Paper 116, London, British Museum Press, 153–164.

Errington S, 1998 *The Death of Authentic Primitive Art and Other Tales of Progress*, Berkley, University of California Press.

Faber P, 1993 Towards a global art museum. In Leyten H and Damen B (eds), *Art, Anthropology and the Modes of Re-presentation*, Amsterdam, Royal Tropical Institute, 61–70.

Fahy A, 1995 *Collections Management*, London and New York, Routledge.

Farnsworth J, 1997 Sculpture in active service. In Lindley P (ed), *Sculpture Conservation, Preservation or Interference?*, Aldershot, Scolar Press, 39–54.

Federspeil B, 1999 Our creative diversity and contemporary issues in conservation. In Bridgland J (ed), *ICOM Committee for Conservation Preprints of the 12th Triennial Meeting Lyon 29 August – 3 September 1999*, London, James & James, 166–171.

Fell V, 1996 Washing away the evidence. In Roy A and Smith P (eds), *Archaeological Conservation and its Consequences. Preprints of the Contributions to the Copenhagen Congress*, London, IIC, 48–51.

Feller R, 1989 Concerning the place of science in the scheme of things. In Baer N S (ed), *Training in Conservation*, New York, New York University Institute of Fine Arts, 17–32.

Feller R, 1994 Aspects of chemical research in conservation: the deterioration process, *Journal of the American Institute for Conservation* 33, 91–99.

Fenn J, 1995 The cellulose nitrate time bomb: using sulphonophthalein indicators to evaluate storage strategies. In Heuman J (ed), *From Marble to Chocolate, the Conservation of Modern Sculpture, Tate Gallery Conference September 1995*, London, Archetype, 87–92.

Ferguson R, 1992 Trends in conservation education: war of words or words of war?, *International Journal of Museum Management and Curatorship* 11, 411–418.

Fforde C, 1991 The preservation of cultural property, *World Archaeological Bulletin* 5, 30–42.

Fforde C, 1992 English collections of human remains: an introduction, *World Archaeological Bulletin* 6, 1–10.

Fielden B M, 1987 *Between Two Earthquakes, Cultural Property in Seismic Zones*, Los Angeles, Getty Conservation Institute.

Fitch J M, 1982 (fourth reprinting 1998) *Historic Preservation: Curatorial Management of the Built World*, Charlottesville and London, University Press of Virginia.

Fleming S J, 1975 *Authenticity in Art: the Scientific Detection of Forgery*, London, Institute of Physics.

Fletcher M, 1988 V&A admits 'disastrous failure', *The Times* 15th December 1988.

Florian M-L, 1990 *The Conservation of Artifacts made from Plant Materials*, Los Angeles, Getty Conservation Institute.

Foley K, 1989 *Education and Training for Conservation. A Report for the Conservation Unit of the Museums and Galleries Commission*, London, Museums and Galleries Commission.

Foley K, 1992 Education and training for practice. In White R (ed), *Archaeological Conservation, Training and Employment. Papers presented at a meeting organised by the Archaeology Section of UKIC, May 1991*, London, UKIC, 27–28.

Foley K, 1999a FULCO, a framework of competence for professional conservation-restoration practice: the project discussed. In Bridgland J (ed), *ICOM Committee for Conservation Preprints of the 12th Triennial Meeting Lyon 29 August–3 September 1999*, London, James & James, 139–146.

Foley K, (ed) 1999b *Conservation Future Challenges. Proceedings of a meeting at West Dean College, November 1998* London, English Heritage.

Foley K and Scholten S, 1998 *FULCO - a Framework of Competence for Conservator-Restorers in Europe. A discussion paper for the Vienna Meeting 30 November – 1 December 1998*, Amsterdam, Instituut Collectie Nederland.

Forbes R J, 1955 onwards *Studies in Ancient Technology* 11 vols, Leiden, E J Brill.

Forrester J, 1997 Coming out, *Museums Journal* June 1997, 34.

France-Lanord A, 1996 Knowing how to 'question' the object before restoring it. In Stanley-Price N, Talley M K and Vaccaro A M (eds), *Historical and Philosophical Issues in the Conservation of Cultural Heritage*, Los Angeles, The Getty Conservation Institute, 244–247.

Fraser D, 1997 Conservation area, *Museums Journal* June 1997, 31–33.

Fry M F, 1996 Buried but not forgotten: sensitivity in disposing of major archaeological timbers. In Roy A and Smith P (eds), *Archaeological Conservation and its Consequences, Preprints of the Contributions to the Copenhagen Congress*, London, IIC, 52–54.

Fry R, 1997 Preservation or desecration? The legal position of the restorer. In Lindley P (ed), *Sculpture Conservation, Preservation or Interference?*, Aldershot, Scolar Press, 55–62.

Ganiaris H and Starling K, 1997 Up to standard: planning the needs of an archaeological archive. In Roy A and Smith P (eds), *Archaeological Conservation and its Consequences. Preprints of the Contributions to the Copenhagen Congress*, London, IIC, 55–58.

Ganiaris H and Sully D, 1998 Showcase construction: materials and methods used at the Museum of London, *The Conservator* 22, 57–67.

Gathercole P, 1989 The fetishism of artifacts. In Pearce S (ed), *Museum Studies in Material Culture*, London and New York, Leicester University Press, 73–82.

Gathercole P (Chair), 1995 Where would you draw the line between ephemera you would conserve and that which you would let die? In Hill L and Giles S (eds), *Where to Start and Where to Stop?* MEG Occasional Papers 4, Hull, Museum Ethnographers Group, 30–31.

Gedye I and Hodges H, 1964 The teaching of archaeological conservation, *Bulletin of the Institute of Archaeology* 4, 83–87.

Geijer A and Franzen A M, 1975 Textile conservation in Sweden: problems and practice. In Leigh D, Moncrieff A, Oddy A and Pratt P (eds), *Conservation in Archaeology and the Applied Arts. 1975 Stockholm Congress*, London, International Institute for Conservation, 7–13.

Gilberg M, 1987 Friedrich Rathgen: the father of modern archaeological conservation, *Journal of the American Institute of Conservation* 26, 105–120.

Gilberg M, 1988a Antonio Piaggio and the conservation of the Herculaneum papyri. In Daniels V (ed), *Early Advances in Conservation*, British Museum Occasional Paper 65, London, British Museum, 1–6.

Gilberg M, 1988b History of bronze disease and its treatment. In Daniels V (ed), *Early Advances in Conservation*, British Museum Occasional Paper 65, London, British Museum, 59–70.

Gilberg M, 1997 Alfred Lucas: Egypt's Sherlock Holmes, *Journal of the American Institute of Conservation* 36, 31–48.

Giles M, 1989 *Conservation Training, An Initial Survey of Full-time Courses in the United Kingdom*, London, Museums and Galleries Commission.

Gillard R D, Hardman S M, Thomas R G and Watkinson D E, 1994 The mineralisation of fibres in burial environments, *Studies in Conservation* 39, 132–140.

Ginell W S, 1994 The nature of changes caused by physical factors. In Krumbein W E, Brimblecombe P, Cosgrove D E and Staniforth S (eds), *Durability and Change: The Science, Responsibility, and Cost of Sustaining Cultural Heritage. Report of the Dahlem Workshop, December 1992*, Chichester, John Wiley, 81–94.

Giusti A, 1999 Reversibility in the restoration of stone artifacts: real possibilities and objective limits. In Oddy A and Carroll S (eds), *Reversibility – does it exist?* British Museum Occasional Paper 135, London, British Museum, 91–98.

Glaser J R and Zenetou A A, 1996 *Museums, A Place to Work*, London and New York, Routledge, in association with the Smithsonian Institution.

Godonou A, 1995 PREMA, training programme for African museums. In Hill L and Giles S (eds). *Where to Start and Where to Stop?* MEG Occasional Papers 4, Hull, Museum Ethnographers Group, 54–56.

Goldberg L, 1996 A history of pest control measures in the anthropology collections, *Journal of the American Institute for Conservation* 35, 23–42.

Goodburn-Brown D, 1996 The effect of various conservation treatments on the surface of metals from waterlogged sites, in London. In Roy A and Smith P (eds), *Archaeological Conservation and its Consequences, Preprints of the Contributions to the Copenhagen Congress*, London, IIC, 59–64.

Green L R and Thickett D, 1995 Testing materials for use in storage and display of antiquities: a revised methodology, *Studies in Conservation* 40, 145–152.

Greene V, 1992 'Accessories of holiness': defining Jewish sacred objects, *Journal of the American Institute for Conservation* 31, 31–39.

Greenfield J, 1995 *The Return of Cultural Treasures*, Cambridge, Cambridge University Press.

Greenhill B, 1976 *Archaeology of the Boat, A New Introductory Study*, London, Adam and Charles Black.

Greeves T, 1989 Archaeology and the green movement: a case for *perestroika*, *Antiquity* 63, 659–666.

de Guichen G, 1982 A challenge to the profession, *Museum* 34(1), 4–5.

de Guichen G, 1984 Object interred, object disinterred. In Stanley-Price N (ed), *Conservation of Archaeological Excavations*, Rome, ICCROM, 21–30.

de Guichen G, 1988 *Climate in Museums*, 2nd edition, Rome, ICCROM.

de Guichen G, 1991 Scientists and the preservation of cultural heritage. In Baer N S, Sabbioni C and Sors A I (eds), *Science, Technology and European Cultural Heritage. Proceedings of the European Symposium, Bologna, Italy, 13-16 June 1989*, London, Butterworth-Heinemann, 17–25.

de Guichen G, 1999 Preventive conservation: a mere fad or far-reaching change? *Museum International* 51(1), 4–6.

Hamilton A, 1999 Heritage £1m keeps castle in ruins, *The Times*, October 13.

Hammil J and Cruz R, 1989 Statement of American Indians against desecration before the World Archaeological Congress. In Layton R (ed), *Conflict in the Archaeology of Living Traditions*, London, Routledge, 195–200.

Hanington D, 2000 To treat or not to treat – that is the question, *CCI Newsletter* 25, 1–3.

Hanna S, 1990 The cleaning and removal of surface coatings from a seventh-century BC sandstone shrine from Nubia. In Mills J S and Smith P (eds), *Cleaning, Retouching and Coatings, Technology and Practice for Easel Paintings and Polychrome Sculpture. Preprints of the Contributions to the Brussels Congress*, London, IIC, 23–27.

Hanna S and Lee N J, 1988 The consequences of previous adhesives and consolidants used for stone conservation at the British Museum. In Daniels V (ed), *Early Advances in Conservation*, British Museum Occasional Paper 65, London, British Museum, 89–102.

Hansen E F, Walston S and Bishop M H (eds), 1993 *Matte Paint. Its History and Technology, Analysis, Properties and Conservation Treatment with Special Emphasis on Ethnographic Objects*, Marina del Rey, Getty Conservation Institute and IIC.

Harding E and Oddy A, 1992 Leonardo da Vinci's Cartoon *The Virgin and Child with St Anne and St John the Baptist*. In Oddy A (ed), *The Art of the Conservator*, London, British Museum Press, 28–41.

Harrison R, (ed) 1994 *The Manual of Heritage Management*, Oxford, Butterworth-Heinemann.

Hather J, 2000 *The Identification of the Northern European Woods, A Guide for Archaeologists and Conservators*, London, Archetype.

Hatton A, 1994 The legislation and institutional context: museums. In Harrison R (ed), *Manual of Heritage Management*, Oxford, Butterworth-Heinemann, in association with the Association of Independent Museums, 157–170.

Hawkins J and Macdonald S, 1997 Roofs of England: reviving a lost industry, *Conservation Bulletin* July 1997, 10.

Hebditch M, 1998 Conservation research in museums. In Foley K and Shacklock V (eds), *Conservation Research, Needs and Provision. Proceedings of the Colloquium held on 1 July 1997 at De Montfort University Leicester sponsored by the Museums and Galleries Commission*, Leicester, De Montfort University, 9–14.

Hedley G, 1986 Cleaning and meaning, *The Ravished Image* reviewed, *The Conservator* 10, 2–6.

Hedley G, 1990 Long-lost relations and new found relativities: issues in the cleaning of paintings. In Todd V (ed), *Appearance, Opinion, Change, Evaluating the Look of Paintings*, London, UKIC, 8–13.

Heritage Preservation and National Parks Service, 1998 *Caring for Your Historic House*, New York, Harry N Abrams Inc.

Heuman J, 1997 Perspectives on the repatination of outdoor bronze sculptures. In Lindley P (ed), *Sculpture Conservation, Preservation or Interference?*, Aldershot, Scolar Press, 121–128.

Heuman J, 1999a Introduction. In Heuman J (ed), *Material Matters, The Conservation of Modern Sculpture*, London, Tate Gallery Publishing, 7–13.

Heuman J, 1999b OTTOshaft 1992. In Heuman J (ed), *Material Matters, The Conservation of Modern Sculpture*, London, Tate Gallery Publishing, 90–99.

Hewison R, 1987 *The Heritage Industry*, London, Methuen.

Heywood F, 2000 Going it alone, *Museums Journal* February 2000, 53.

Hijara I, 1980 Arpachiyah 1976, *Iraq* 42.

Hill C C, 1990 Museum curator: roles and responsibilities. In Ramsay-Jolicoeur B A and Wainwright I N M (eds), *Shared Responsibility. Proceedings of a Seminar for Curators and Conservators, co-hosted by the National Gallery of Canada and the Canadian Conservation Institute, Department of Communications, October 1989*, Ottawa, National Gallery of Canada, 19–23.

Hinton P, 1999 Archaeology: a professional approach. In *IFA Yearbook and Directory of Members 1999*, Reading, IFA, 16–17.

HLF (Heritage Lottery Fund), 1998 *Conservation Plans for Historic Places*, London, Heritage Lottery Fund.

Hoadley R B, 1978 The dimensional response of wood to variations in relative humidity. In Brommelle N S, Moncrieff A and Smith P (eds), *Conservation of Wood in Painting and the Decorative Arts. Preprints of the Contributions to the Oxford Congress, September 1978*, London, IIC, 1–6.

Hodder I, 1989 Writing archaeology: site reports in context, *Antiquity* 63, 268–74.

Hodder I, 1992 *Theory and Practice in Archaeology*, London, Routledge.

Hodder I, Shanks M, Alexandri A, Buchli V, Carman J, Last J and Lucas G (eds), 1995 *Interpreting Archaeology, Finding Meaning in the Past*, London and New York, Routledge.

Hodges H, 1963 An *ab initio* course in the conservation of antiquities, In Thomson G (ed), *Recent Advances in Conservation*, London, Butterworths, 211–214.

Hodges H W M, 1964 (1989) *Artifacts*, London, John Baker,

Hodges H W M, 1970 *Technology in the Ancient World*, London, Allen Lane, The Penguin Press.

Hodkinson I S, 1990 Man's effect on paintings. In Ramsay-Jolicoeur B A and Wainwright I N M (eds), *Shared Responsibility, Proceedings of a Seminar for Curators and Conservators, National Gallery of Canada, 26th-28th October 1998*, Ottawa, National Gallery of Canada, 54–68.

Holben-Ellis M, 1990 The Conservation Training Program, Institute of Fine Arts, New York University: past and future directions. In Stanley-Price N (ed), *The Graduate Conservator in Employment, Expectations and Realities. Proceedings of the Interim Meeting of the Working Group on Training in Conservation and Restoration of the ICOM Committee of Conservation, Amsterdam and Leiden 31 August – 1 September 1989*, Amsterdam, Oplieding Restauratoren, 40–49.

Holland M, 1825 *The Modern Family Receipt Book*, London, Thomas Tegg.

Holtorf C and Schadla-Hall T, 1999 Age as artifact: on archaeological authenticity, *European Journal of Archaeology* 2, 229–247.

Hooper-Greenhill E, 1992 *Museums and the Shaping of Knowledge*, London, Routledge.

Hooper-Greenhill E, 1997 *Cultural Diversity, Developing Museum Audiences in Britain*, London and New York, Leicester University Press.

Hooper-Greenhill E, 1998 Perspectives on Hinemihi, a Maori meeting house. In Barringer T and Flynn T (eds), *Colonialism and the Object, Empire, Material Culture and the Museum*, London and New York, Routledge, 129–143.

Horie C V, 1987 *Materials for Conservation*, Oxford, Butterworth-Heinemann and Architectural Press.

Howie F M P, 1984 Materials used for preserving fossil specimens since 1930: a review. In Brommelle N S, Pye E M, Smith P and Thomson G (eds), *Adhesives and Consolidants. Preprints of the Contributions to the Paris Congress September 1984*, London, International Institute for Conservation, 92–97.

Hubert J, 1989a First World Archaeological Congress Inter-Congress, Vermillion, South Dakota, USA, *World Archaeological Bulletin* 4, 14–19.

Hubert J 1989b A proper place for the dead: a critical review of the reburial issue. In Layton R (ed), *Conflict in the Archaeology of Living Traditions*, London, Routledge, 131–166.

Hughes R, 1993 Artificial patination. In La Niece S and Craddock P (eds), *Metal Plating and Patination*, Oxford, Butterworth-Heinemann, 1–18.

Humphrey J W, Oleson J P and Sherwood A N, 1998 *Greek and Roman Technology, A Sourcebook*, London and New York, Routledge.

Hunter K and Foley K, 1984 The Lincoln Hanging Bowl. In Bacon L and Knight B (eds), *From Pinheads to Hanging Bowls*, London, UKIC Occasional Papers 7,16–18.

ICOM (International Council for Museums), 1969 *Problems of Conservation in Museums. A selection of papers presented to the joint meeting of the ICOM Committee for Museum Laboratories and the ICOM Committee for the Care of Paintings, held in Washington and New York, September 1965*, Paris, Editions Eyrolles, and London, George Allen and Unwin.

ICOM (International Council for Museums), 1996 *Statutes and Code of Professional Ethics*, Paris, ICOM.

ICOM-Conservation Committee (ICOM-CC), 1984 *The Conservator-Restorer: A Definition of the Profession.*

ICOMOS, 1964 *International Charter for the Conservation and Restoration of Monuments and Sites (the Venice Charter)*, International Council for Monuments and Sites (http://www.icomos.org/docs/venice_charter.html)

ICOMOS New Zealand, 1992 *Charter for the Conservation of Places of Cultural Heritage Value* (http://www.icomos.org/docs/nz_92charter.html)

Idiens D, 1995 Museum responsibilities II. In Hill L and Giles S (eds), *Where to Start, Where to Stop. Papers from the British Museum / MEG Ethnographic Conservation Colloquium*, Museum Ethnographers Group Occasional Papers 4, Hull, Museum Ethnographers Group, 35–36.

IFA (Institute of Field Archaeologists), 1999 Code of conduct. In *IFA Yearbook and Directory of Members 1999*, Reading, IFA, 9–13.

IIC, 1997 Historical note. In *The International Institute for the Conservation of Historic and Artistic Works, List of Members 1997–8*, London, IIC, 3–4.

IIC, 1999 Memorandum of Association (1950). In *The International Institute for the Conservation of Historic and Artistic Works, List of Members 1999–2000*, London, IIC, 5–8.

IIC-CG, 1989 *Code of Ethics and Guidance for Practice*, Ottawa, International Institute for Conservation – Canadian Group.

Impey O and MacGregor A, (eds) 1985 *The Origins of Museums*, Oxford, Oxford University Press.

Ito N, 1995 'Authenticity' inherent in cultural heritage in Asia and Japan. In Larsen K E (ed), *Nara Conference on Authenticity in Relation to the World Heritage Convention, November 1994*, Tokyo, UNESCO World Heritage Centre; Agency for Cultural Affairs (Japan); ICCROM; ICOMOS, 35–46.

Jacobsen T, 1988 Iron corrosion theories and the conservation of archaeological iron objects in the 19th century, with an emphasis on Scandinavian and German sources. In Daniels V (ed), *Early Advances in Conservation*, British Museum Occasional Paper 65, London, British Museum, 51–58.

Jaeschke H F, 1996 The conservation treatment of looted antiquities and the responsibility of conservators. In Roy A and Smith P (eds), *Archaeological Conservation and its Consequences. Preprints of the Contributions to the Copenhagen Congress*, London, IIC, 82–85.

Jagger N and Aston J, 1999 *Nurturing Conservators, the Early Career Paths of Conservation Graduates*, London, Institute for Employment Studies.

Janaway R, 1989 Corrosion preserved textile evidence: mechanism, bias and interpretation. In Janaway R and Scott B (eds), *Evidence Preserved in Corrosion Products, New Fields in Artifact Studies*, UKIC Occasional Papers 8, London, UKIC, 21–29.

Janaway R and Scott B (eds), 1989 *Evidence Preserved in Corrosion Products, New Fields in Artifact Studies*, UKIC Occasional Papers 8, London, UKIC.

Janssen R, 1986 *Egyptian Textiles*, Princes Risborough, Shire Publications.

Jedrzejewska H, 1983 Psychological factors in conservation and museology. In Timar-Balazsy A (ed), *Fourth International Restorer Seminar vol 1, Restorer's – Museologist's-Scientist's Common Responsibility in the Protection of Museum Objects*, Budapest, National Centre of Museums, 31–38.

Jessup G, 1991 *Outcomes, NVQs and the Emerging Model of Education and Training*, London, New York, Philadelphia, The Falmer Press.

Johnson J, 1999 Masked hazard, *Common Ground, Archaeology and Ethnography in the Public Interest,* Fall 1999, 26–31.

Jokilehto J, 1995 Authenticity: a general framework for the concept. In Larsen K E (ed), *Nara Conference on Authenticity in Relation to the World Heritage Convention, November 1994*, Tokyo, UNESCO World Heritage Centre; Agency for Cultural Affairs (Japan); ICCROM; ICOMOS, 17–34.

Jokilehto J, 1999 *A History of Architectural Conservation*, Oxford, Butterworth-Heinemann.

Jones M, (ed) 1990 *Fake? The Art of Deception*, London, British Museum Publications.

Jury L, 2000 Restorers 'are ruining Titians', *The Independent on Sunday*, 27 February.

Katz S, 1984 *Classic Plastics*, London, Thames and Hudson.

Kavanagh G, (ed) 1994 *Museum Provision and Professionalism*, London and New York, Routledge.

Keck S, 1963 Training for engineers in conservation. In Thomson G (ed), 1963 *Recent Advances in Conservation*, London, Butterworths, 199–201.

Keck S, 1976 The life expectancy of materials and problems of increasing visitor use. In International Centre for Conservation, Rome, and International Centre Committee of the Advisory Council on Historic Preservation, *Preservation and Conservation, Principles and Practices. Proceedings of the North American International Regional Conference*, Washington, The Preservation Press, 327–339.

Keck S, 1996 Further materials for a history of conservation. In Stanley-Price N, Talley M K and Vaccaro A M (eds), *Historical and Philosophical Issues in the Conservation of Cultural Heritage*, Los Angeles, The Getty Conservation Institute, 281–287.

Keene S (ed), 1980 *Conservation, Archaeology and Museums*, United Kingdom Institute for Conservation Occasional Paper 1, London, UKIC.

Keene S, 1983 Conservation Records - editorial introduction, *The Conservator* 7, 4.

Keene S, 1987 The Winchester reliquary: conservation and elucidation. In Black J W (ed), *Recent Advances in the Conservation and Analysis of Artifacts*, Institute of Archaeology, Jubilee Conservation Conference Papers, London, Summer Schools Press, 25–32.

Keene S, 1991 Audits of care: a framework for collections condition surveys. In Norman N and Todd V (eds), *Storage. Papers given at a UKIC conference Restoration '91*, London, UKIC, 6–16.

Keene S, 1994 Objects as systems: a new challenge for conservation. In Oddy A (ed), *Restoration - Is It Acceptable?*, British Museum Occasional Papers 99, London, British Museum, 19–26.

Keene S, 1996 *Managing Conservation in Museums*, Oxford, Butterworth-Heinemann.

King J C H, 1999 *First Peoples, First Contacts*, London, British Museum Press.

Kingery D W and Vandiver P B, 1986 *Ceramic Masterpieces*, London and New York, The Free Press (Macmillan).

Kirshenblatt-Gimblett B, 1991 Objects of ethnography. In Karp I and Levine S D (eds), *Exhibiting Cultures, The Poetics and Politics of Museum Display*, Washington and London, Smithsonian Institution Press, 386–443.

Kite M, 1992 Gut membrane, parchment and gelatine incorporated into textile objects, *The Paper Conservator* 16, 98–105.

Knell S (ed), 1994 *Care of Collections*, London and New York, Routledge.

Koehn D, 1994 *The Ground of Professional Ethics*, London and New York, Routledge.

Koestler R J (ed), 1991 *Biodeterioration of Cultural Property*, New York, Elsevier.

Koestler R J (Rapporteur), 1994 Group report: How do external environmental factors accelerate change? In Krumbein W E, Brimblecombe P, Cosgrove D E and Staniforth S (eds), *Durability and Change. The Science, Responsibility, and Cost of Sustaining Cultural Heritage. Report of the Dahlem Workshop, December 1992*, Chichester, John Wiley, 149–164.

Kollandsrud K, 1997 Vasari's theory on the origins of oil painting and its influence on cleaning methods: The ruined polychromy of the early thirteenth-century crucifix from Haug, Norway. In Lindley P (ed), *Sculpture Conservation, Preservation or Interference?*, Aldershot, Scolar Press, 139–150.

Koller M, 1994 Learning from the history of preventive conservation. In Roy A and Smith P (eds), *Preventive Conservation, Practice, Theory and Research. IIC Preprints of the Contributions to the Ottawa Congress, 12-16 September 1994*, London, International Institute for the Conservation of Historic and Artistic Works, 1–7.

Koob S, 1987 Detachable plaster restorations for archaeological ceramics. In Black J (ed), *Recent Advances in the Conservation and Analysis of Artifacts*, London, Summer Schools Press, 63–65.

Koob S, 1998 Obsolete fill materials found on ceramics, *Journal of the American Institute for Conservation* 37, 49–67.

Kopytoff I, 1986 The cultural biography of things: commoditization as process. In Appadurai A (ed), *The Social Life of Things, Commodities in Cultural Perspective*, Cambridge, Cambridge University Press, 64–91.

Kosek J M, 1994 Restoration of art on paper in the West: a consideration of changing attitudes and values. In Oddy A (ed), *Restoration, Is it Acceptable?*, British Museum Occasional Paper 99, London, British Museum, 41–50.

Krumbein W E, Brimblecombe P, Cosgrove D E and Staniforth S (eds), 1994 *Durability and Change. The Science, Responsibility, and Cost of Sustaining Cultural Heritage. Report of the Dahlem Workshop, December 1992*, Chichester, John Wiley.

La Rocca E and Nardi R, 1994 Preventive conservation and restoration: a matter of costs. In Roy A and Smith P (eds), *Preventive Conservation, Practice Theory and Research. Preprints of the Contributions to the Ottawa Congress September 1994*, London, IIC, 24–27.

Labi K A, 1993 The theory and practice of conservation among the Akans of Ghana. In Bridgland J (ed), *Preprints for the 10th Triennial Meeting Washington, DC, USA 22-27 August 1993*, Paris, ICOM-CC, 371–376.

Lambert J B, 1997 *Traces of the Past, Unravelling the Secrets of Archaeology through Chemistry*, Reading, Massachusetts, Addison Wesley.

Landi S, 1985 *The Textile Conservator's Manual*, Oxford, Butterworths.

Landi S and Hall R, 1979 The discovery and conservation of an ancient Egyptian linen tunic, *Studies in Conservation* 24, 141.

Lang J and Middleton A (eds), 1997 *Radiography of Cultural Material* London, Butterworth-Heinemann.

Larsen K E (ed), 1995 *Nara Conference on Authenticity in relation to the World Heritage Convention, November 1994* Tokyo, UNESCO World Heritage Centre; Agency for Cultural Affairs (Japan); ICCROM; ICOMOS.

Larson J, 1990 The treatment and examination of painted surfaces on eighteenth-century terra-cotta sculptures. In Mills J S and Smith P (eds), *Cleaning, Retouching and Coatings, Technology and Practice for Easel Paintings and Polychrome Sculpture. Preprints of the Contributions to the Brussels Congress*, London, IIC, 28–32.

Larson J, 1992 Statue of the Bodhisattva Guanyin. In Oddy A (ed), *The Art of the Conservator*, London, British Museum Press, 177–189.

Larson J, 1997 Sculpture conservation: treatment or reinterpretation? In Lindley P (ed), *Sculpture Conservation, Preservation or Interference?*, Aldershot, Scolar Press, 69–82.

Lasko P, 1992 The principles of restoration. In Kahn D (ed), *The Romanesque Frieze and its Spectator, The Lincoln Symposium Papers*, London, Harvey Miller, for the Dean and Chapter of Lincoln Cathedral, 143–162.

Lasko P and Lodewicks J, 1982 Curator and scientist: towards unity of aim, *Museum* 34(1), 31–33.

Layton R, (ed) 1989a *Who Needs the Past?*, London, Unwin Hyman.

Layton R, (ed) 1989b *Conflict in the Archaeology of Living Traditions*, London and New York, Routledge.

Lazzarini L and Salvadori O, 1989 A reassessment of the formation of the patina called *scialbatura*, *Studies in Conservation* 34, 20–26.

Leggett J, 1999 National consciousness, *Museums Journal*, October 1999, 17.

Leigh D, 1990 New perceptions, new expectations. In Stanley-Price N (ed), *The Graduate Conservator in Employment, Expectations and Realities. Proceedings of the Interim Meeting of the Working Group on Training in Conservation and Restoration of the ICOM Committee of Conservation, Amsterdam and Leiden 31 August – 1 September 1989*, Amsterdam, Oplieding Restauratoren, 111–120.

Leigh D (Rapporteur), 1994 Group report: What are the responsibilities for cultural heritage and where do they lie? In Krumbein W E, Brimblecombe P, Cosgrove D E and Staniforth S (eds), *Durability and Change. The Science, Responsibility, and Cost of Sustaining Cultural Heritage. Report of the Dahlem Workshop, December 1992*, Chichester, John Wiley, 268–286.

Leighton J, 1998 Taking a closer look: art historians, restorers and scientists. In Hermens E (ed), *Looking through Paintings, The Study of Painting Techniques and Materials in Support of Art Historical Research*, Baarn Vitgeverij de Prom and London, Archetype, 29–34.

Lerner S and Hoffman T, 2000 Bringing archaeology to the public: programmes in the Southwestern United States. In McManamon F P and Hatton A (eds), *Cultural Resource Management in Contemporary Society, Perspectives on Managing and Presenting the Past*, London and New York, Routledge, 231–246.

Lewis G, 1992 Museums and their precursors: a brief world survey. In Thompson J M A (ed), *The Manual of Curatorship*, Oxford, Butterworth-Heinemann, 5–21.

Leyten H, 1993 Foreword. In Leyten H and Damen B (eds), *Art, Anthropology and the Modes of Re-presentation*, Amsterdam, Royal Tropical Institute, 7–14.

Lipe W D, 1984 Value and meaning in cultural resources. In Cleere H (ed), *Approaches to the Archaeological Heritage*, Cambridge, Cambridge University Press, 1–11.

Lister D, 1999 Gallery told, some of your art is worthless, *The Independent*, London, 24th June.

Lister D, 2000 Tate Modern's all very well, says MacGregor of the National. But museums are in peril, *The Independent*, 26th May.

Lloyd H and Mullaly T, 1994 The impact of over visiting: methods of assessing the sustainable capacity of historic houses. In Roy A and Smith P (eds), *Preventive Conservation, Practice Theory and Research. Preprints of the Contributions to the Ottawa Congress September 1994*, London, IIC, 132–138.

Lord A, 1996 The conservation of sixteen masonic aprons: treating multi-media ritual artifacts, *The Conservator* 20, 95–103.

Lord B and Lord G D, 1997 *The Manual of Museum Management*, London, The Stationery Office.

Lowenthal D, 1979 Age and artifact: dilemmas of appreciation. In Meinig D W (ed), *The Interpretation of Ordinary Landscapes*, Oxford, Oxford University Press, 103–128.

Lowenthal D, 1985 *The Past is a Foreign Country*, Cambridge, Cambridge University Press.

Lowenthal D, 1994 The value of age and decay. In Krumbein W E, Brimblecombe P, Cosgrove D E and Staniforth S (eds), *Durability and Change. The Science, Responsibility, and Cost of Sustaining Cultural Heritage. Report of the Dahlem Workshop, December 1992*, Chichester, John Wiley, 39–49.

Lowenthal D, 1997 *The Heritage Crusade and the Spoils of History*, Cambridge, Cambridge University Press.

Lowenthal D and Binney M (eds), 1981 *Our Past Before Us, Why Do We Save It?*, London, Temple Smith.

Maddison D and Mourato S, 1999 *Valuing Different Road Options for Stonehenge*, London, University College London Centre for Social and Economic Research on the Global Environment.

Magor L, 1990 An artist's thoughts on conservation and curatorial issues. In Ramsay-Jolicoeur B A and Wainwright I N M (eds), *Shared Responsibility, Proceedings of a Seminar for Curators and Conservators; October 1989*, National Gallery of Canada and the Canadian Conservation Institute, Department of Communications, 6–11.

Maish J P, 1995 Interpretation of ancient artistry: conservation of a gold wreath from the fourth century BC, *Journal of the American Institute for Conservation* 34, 113–127.

Mancinelli F, 1991 The frescoes of Michelangelo on the vault of the Sistine Chapel: conservation methodology, problems and results. In Cather S (ed), *The Conservation of Wall Paintings. Proceedings of a symposium organised by the Courtauld Institute of Art and the Getty Conservation Institute, London, July 13-17, 1987*, Los Angeles, The Getty Conservation Institute, 57–66.

Mancinelli F, 1992 Michelangelo's frescoes in the Sistine Chapel. In Oddy A (ed), *The Art of the Conservator*, London, British Museum Press, 89–107.

Mann P R, 1994a Working exhibits and the destruction of evidence in the Science Museum. In Knell S (ed), *Care of Collections*, London, Routledge, 35–46.

Mann P R, 1994b The restoration of vehicles for use in research, exhibition, and demonstration. In Oddy A (ed), *Restoration, Is it Acceptable?*, British Museum Occasional Paper 99, London, British Museum, 131–138.

Marcon P, 1997 Decision support models for preventive conservation. In Bradley S (ed), *The Interface between Science and Conservation*, British Museum Occasional Paper 116, London, British Museum, 143–152.

Marijnissen R H, 1996 Degradation, conservation, and restoration of works of art: historical overview. In Stanley-Price N, Talley M K and Vaccaro A M (eds), *Historical and Philosophical Issues in the Conservation of Cultural Heritage*, Los Angeles, The Getty Conservation Institute, 275–280.

Marks R and Blench B J R, 1979 *The Warwick Vase*, Glasgow, The Burrell Collection; Glasgow Museums and Art Galleries.

Martin-Gil J, Ramos-Sanchez M C and Martin-Gil F J, 1999 Ancient pastes for stone protection against environmental agents, *Studies in Conservation* 44, 58–62.

Massing A, 1998 Restoration policy in France in the eighteenth century. In Sitwell C and Staniforth S (eds). *Studies in the History of Painting Restoration*, London, Archetype Publications in association with the National Trust, 63–84.

Matero F, 1993 The conservation of immovable cultural property: ethical and practical dilemmas, *Journal of the American Institute for Conservation* 32, 15–21.

Mathisen S A and Rushfield R A, 1999 What do students expect from Internships. In Dardes K (ed), *Defining and Measuring Effectiveness in Education and Training. Proceedings of the interim meeting of the ICOM-CC Working Group on Training in Conservation and Restoration, Vantaa, Finland*, Vantaa, Evtek Institute of Art and Design, 55–70.

Mattson E, Nord A G, Tronner K, Fjaestad M, Lagerlöf A, Ullén I and Borg G Ch, 1996 *Deterioration of Archaeological Material in Soil*, Results on *Bronze Artifacts Rapport RIK 10*, Stockholm, The Central Board of National Antiquities and the National Historical Museums.

McBryde I, 1985 Introduction. In McBryde I (ed), *Who Owns the Past?*, Melbourne, Oxford University Press, 1–10.

McCarthy M, 2000 Men from the ministry dismiss sparrows as vermin, *The Independent*, 10th June.

McClure I, 1992 Henry, Prince of Wales, on Horseback by Robert Peake the Elder. In Oddy A (ed), *The Art of the Conservator*, London, British Museum Press, 59–72.

McCrady E R, 1997 Can scientists and conservators work together? In Bradley S (ed), *The Interface between Science and Conservation*, British Museum Occasional Paper 116, London, The British Museum, 243–247.

McCredie A, 1991 Relationships between conservators' codes of ethics and the conservation policies of institutions – common threads, conflicts and failings, *New Zealand Museums Journal* 21, 6–9.

McKee P, 1999 A story of storage, *Museums Journal* January 2000, 14.

McLaren D, 1999 Mark Dion (review), *The Independent*, 12th December.

McLeod M D, 1995 Harold Gowers Memorial Lecture 1989 'Where to start, where to stop'. In Hill L and Giles S (eds), *Where to Start, Where to Stop. Papers from the British Museum / MEG Ethnographic Conservation Colloquium*, Museum Ethnographers Group Occasional Papers 4, Hull, Museum Ethnographers Group, 5–9.

McManamon F P and Hatton A, 2000 Introduction: considering cultural resource management in modern society. In McManamon F P and Hatton A (eds), *Cultural Resource Management in Contemporary Society*, London and New York, Routledge, 1–19.

McManamon F P and Morton S D, 2000 Reducing the illegal trafficking in antiquities, In McManamon F P and Hatton A (eds), *Cultural Resource Management in Contemporary Society*, London and New York, Routledge, 247–275.

Meehan P, Buck P, and Lee L, 1996 The investigation and conservation of a 17th century watch retrieved from the River Thames. *The Conservator* 20, 45–52.

Mellor S, 1992 The exhibition and conservation of African objects: considering the non-tangible, *Journal of the American Institute for Conservation* 31, 3–16.

Mendick R, 2000 Is the Gallery's £40m Rubens just a copy?, *The Independent*, 21st May.

Merrill R, 1990 Conservation in institutions today: how it differs from the past. In Ramsay-Jolicoeur B and Wainwright I N M (eds), *Shared Responsibility, A Seminar for Curators and Conservators*, Ottawa, National Gallery of Canada, 170–173.

Merriman N, 1991 *Beyond the Glass Case*, Leicester, Leicester University Press.

Merriman N, 1996 Understanding Heritage, *Journal of Material Culture* 1(3), 377–386.

Merriman N, 1997 A historical background: how did we get here? Text of a talk given during the session entitled *Specialist Curators – An Endangered Species*? at the Museums Association Conference, Cardiff, 1997 (unpublished).

Merriman N, 2000 The crisis of representation in archaeological museums. In McManamon F P and Hatton A (eds), *Cultural Resource Management in Contemporary Society*, London and New York, Routledge, 300–309.

Merriman N and Swain H, 1999 Archaeological archives: serving the public interest?, *European Journal of Archaeology* 2, 249–267.

Meyers P, 1988 The structure of works of art and historic artifacts. In Sayre E, Vandiver P B, Drusik J and Stevenson C (eds), *Materials Issues in Art and Archaeology I*, Pittsburgh, Pennsylvania, Materials Research Society, 5–18.

MGC (Museums and Galleries Commission), 1992 onwards *Standards in the Museum Care of Collections*, London, Museum & Galleries Commission.

MGC (Museums and Galleries Commission), 1997 *Ours for Keeps?* A resource pack for raising awareness of conservation and collection care London, Museums and Galleries Commission.

Michalski S, 1990a Time's effect on paintings. In Ramsay-Jolicoeur B A and Wainwright I N M (eds), *Shared Responsibility, Proceedings of a Seminar for Curators and Conservators; October 1989*, National Gallery of Canada and the Canadian Conservation Institute, Department of Communications, 39–53,

Michalski S, 1990b Towards specific lighting guidelines. In *9th Triennial Meeting Dresden, German Democratic Republic, August 1990*, Los Angeles, ICOM-CC, 583–588.

Michalski S, 1994 A systematic approach to preservation: description and integration with other museum activities. In Roy A and Smith P (eds), *Preventive Conservation, Practice Theory and Research. Preprints of the Contributions to the Ottawa Congress September 1994*, London, IIC, 8–11.

Middleton V, 1994 Vision, strategy and corporate planning: an overview. In Harrison R (ed), *Manual of Heritage Management*, Oxford, Butterworth-Heinemann in association with the Association of Independent Museums, 3–11.

Miles D, 2000 Holme timber circle: excavation, removal and scientific analysis, *Conservation Bulletin* 37, 6–7.

Miller D, 1994 *Material Culture and Mass Consumption*, Oxford, Blackwell.

Miller E, 1992 The Piranesi Vase. In Oddy A (ed), *The Art of the Conservator*, London, British Museum Press, 122–136.

Mills J S and Smith P (eds), 1990 *Cleaning, Retouching and Coatings, Technology and Practice for Easel Paintings and Polychrome Sculpture. Preprints of the Contributions to the Brussels Congress*, London, IIC.

Milne G, Bates M and Webber M, 1997 Problems, potential and partial solutions: an archaeological study of the tidal Thames, England, *World Archaeology* 29 (1), 130–146.

Mohr E and Schmidt J, 1997 Aspects of economic valution of cultural heritage. In Baer N S and Snethlage R (eds), *Saving Our Architectural Heritage, the Conservation of Historic Stone Structures. Report of the Dahlem Workshop, Berlin, March 1996*, Chichester, John Wiley, 333–348.

Molina T and Pincevin M, 1994 Restoration: acceptable to whom? In Oddy A (ed), *Restoration, Is it Acceptable?*, British Museum Occasional Paper 99, London, British Museum, 77–84.

Moore B P, 1996 The conservation assessment survey as a long-range planning tool: a case study at the Peabody Museum. In *ICOM Committee for Conservation, Preprints of the 11th Triennial Meeting, Edinburgh, September 1996*, London, James & James, 784–792.

Moormann E, 1991 Destruction and restoration of Campanian mural paintings in the eighteenth and nineteenth centuries. In Cather S (ed), *The Conservation of Wall Paintings*, Los Angeles, The Getty Conservation Institute, 87–102.

Mora P, Mora L and Philippot P, 1984 *Conservation of Wall Paintings*, London, Butterworths.

Morgan H and Cruikshank P, 1995 The conservation of the shroud of Resti – an 18th Dynasty linen *Book of the Dead*, In Brown C, Macalister F and Wright M (eds), *Conservation in Ancient Egyptian Collections*. London, Archetype Publications, 1–12.

Morin R, 1999 Creative preservation: the development of an artistic approach to the preservation and presentation of the past, *Conservation and Management of Archaeological Sites* 3, 191–201.

Morris R, 2000 On the Heritage Strategy Review, *Conservation Bulletin* 37, 2–5.

Morrison L, 1986 The conservation of seal-gut parkas, *The Conservator* 10, 17–24.

MTI, 1995 *National Vocational Qualifications and Scottish Vocational Qualifications for the Museums, Galleries and Heritage Community*, Bradford, Museum Training Institute.

Murdoch J, 1994 Defining curation. In Kavanagh G (ed), *Museum Provision and Professionalism*, London and New York, Routledge, 142–146.

Murray Pease, 1964 The Murray Pease Report, *Studies in Conservation* 9, 116–121.

Museums Journal, 1997 Market leaders: qualities at the heart of the best museums were on show at this year's Gulbenkian Awards for Museums and Galleries, *Museums Journal* December 1997, 33.

Museums Journal, 1999 News, views, people and events, *Museums Journal* May 1999, 51.

NAGPRA, 1990 *Native American Graves Protection and Repatriation Act*. USA.

Nardi R, 1995 Open-heart restoration: raising the awareness of the public, *ICOM Committee for Conservation Study Series* 1, 9–11.

Nardi R, 1999 Going public: a new approach to conservation education, *Museum International* 51(1), 44–50.

National Committee to Save America's Cultural Collections, 1992 *Caring for Your Collections*, New York, Harry N Abrams Inc.

National Historic Preservation Act 1966. USA.

National Trust, 2000 Information about the UK National Trust at www.nationaltrust.org.uk (accessed 4th June 2000).

National Trust for Historic Preservation, 1976 *Preservation and Conservation, Principles and Practices. Proceedings of the North American International Regional Conference, conducted under the auspices of the International Centre for Conservation, Rome, Italy and the International Centre Committee of the Advisory Council on Historic Preservation, September 1972*, Washington, The Preservation Press.

Needham J, 1954 onwards *Science and Civilisation in China* (seven volumes), Cambridge, Cambridge University Press.

Nemiroff D, 1990 Copyright: a museum curator's point of view. In Ramsay-Jolicoeur B A and Wainwright I N M (eds), *Shared Responsibility, Proceedings of a Seminar for Curators and Conservators, National Gallery of Canada, 26th-28th October 1998*, Ottawa, National Gallery of Canada, 145–148.

Newbery E and Fecher S, 1994 *In the Nick of Time*, London, Museums and Galleries Commission.

Newey H, Bradley S and Leese M, 1993 Assessing the condition of archaeological iron: an intercomparison. In Bridgland J (ed), *Preprints for the 10th Triennial Meeting Washington, DC, USA 22-27 August 1993*, Paris, ICOM-CC, 786–791

Newey H and Meehan P, 1999 The conservation of the 1895 Panhard et Levassor and the 1922 prototype Austin Seven motorcars: new approaches in the preservation of vehicles, *The Conservator* 2, 11–21.

Newton R and Davison S, 1989 *Conservation of Glass*, London, Butterworths.

NIC, 1984a *Ethnographic and Archaeological Conservation in the United States*, Washington DC, National Institute for the Conservation of Cultural Property, Inc.

NIC, 1984b *A Suggested Curriculum for Training in Ethnographic and Archaeological Conservation*, Washington DC, National Institute for the Conservation of Cultural Property, Inc.

NIC, 1990 *Training for Collections Care and Maintenance, A Suggested Curriculum. Vol 1, Archaeology and Ethnography*, Washington DC, National Institute for the Conservation of Cultural Property, Inc.

Nicholson P T and Shaw I (eds), 2000 *Ancient Egyptian Materials and Technology*, Cambridge, Cambridge University Press.

Nightingale J and Thomas R, 1998 Conservators' jobs go as restructure begins, *Museums Journal* January 1998, 9.

Oakley V, 1990 Vessel glass deterioration at the Victoria and Albert Museum: surveying the collection, *The Conservator* 14, 30–36.

Oddy A (ed), 1992a *The Art of the Conservator*, London, British Museum Press.

Oddy A, 1992b Introduction. In Oddy A (ed), *The Art of the Conservator*, London, British Museum Press, 7–27.

Oddy A, 1997 Obituary, Harold James Plenderleith, *International Institute for Conservation Bulletin* December 1997, 1–2.

Oddy A and Carroll S (eds), 1999 *Reversibility – Does it Exist?*, British Museum Occasional Paper 135, London, British Museum.

Ogden J, 1993 Aesthetic and technical considerations regarding the colour and texture of ancient gold-work. In La Niece S and Craddock P (eds), *Metal Plating and Patination*, Oxford, Butter-worth-Heinemann, 39–49.

van Oosten T B, 1994 Investigation into the degradation of weighted silk. In Verschoor H and Mosk J (eds), *Contributions of the Central Research Laboratory to the Field of Conservation and Restoration*, Amsterdam, Central Research Laboratory for Objects of Art, 65–76.

Organ R, 1961 The conservation of cuneiform tablets, *British Museum Quarterly* 23, 52–57.

Organ R, 1981 Science for the conservator and curator, *Journal of the American Institute for Conservation* 20, 41–44.

Organ R, 1989 A survey of current treatments: a contribution towards a knowledge base for a computer's expert system. *International Restorer Seminar, Vezprem, Hungary*, National Centre of Museums, Budapest, 176–179.

Orna M V (Rapporteur), 1994 Group report: What is durability in artifacts and what inherent factors determine it? In Krumbein W E, Brimblecombe P, Cosgrove D E and Staniforth S (eds), *Durability and Change. The Science, Responsibility, and Cost of Sustaining Cultural Heritage. Report of the Dahlem Workshop, December 1992*, Chichester, John Wiley, 50–66.

Page A E, 1999 Dégradations historiques, *Nike Bulletin*, 4 December 1999, 18–20.

Paine S, 1993 The effects of bat excreta on wall paintings, *The Conservator* 17, 3–10.

Paine S, 1996 *Bats in Churches* Guidelines drawn up for English Heritage in conjunction with English Nature, London, English Heritage.

Palmer N, 1989 Museums and cultural property. In Vergo P (ed). *The New Museology*, London, Reaktion Books, 172–204.

Paterakis A B, 1996 Conservation: preservation versus analysis. In Roy A and Smith P (eds), *Archaeological Conservation and its Consequences, Preprints of the Contributions to the Copenhagen Congress*, London, IIC, 143–149.

Paterson C and Greenfield J, 1998 Storage considerations for native arts: a joint project between Denver Art Museum and the Native American Community. In Wright M and Player-Dahnsjö Y M T (eds), *Site Effects, The Impact of Location on Conservation Treatments*, Edinburgh, The Scottish Society for Conservation and Restoration, 40–45.

Pavelka K L, 1999 Access as a factor contributing to compromise in conservation-treatment decisions. In Oddy A and Carroll S (eds), *Reversibility – Does it Exist?*, British Museum Occasional Paper 135, London, British Museum, 105–110.

Payton R (ed), 1992 *Retrieval of Objects from Archaeological Sites*, London, Archetype.

Peacock E and Turner-Walker G, 1998 The *in situ* conservation of the archbishop's mint workshop, Trondheim, Norway. In Wright M and Player-Dahnsjö Y M T (eds), *Site Effects, The Impact of Location on Conservation Treatments*, Edinburgh, The Scottish Society for Conservation and Restoration, 87–92.

Pearce S, 1990 *Archaeological Curatorship*, Leicester, Leicester University Press.

Pearce S, 1992 *Museums, Objects and Collections*, Leicester, Leicester University Press.

Pearce S, 1995 *On Collecting*, London and New York, Routledge.

Pearce S, 2000 The making of cultural heritage. In Avrami E, Mason R and de la Torre M (eds), *Values and Heritage Conservation*, Los Angeles, The Getty Conservation Institute, 59–64.

Pearson C (ed), 1987 *Conservation of Marine Archaeological Objects*, London, Butterworth.

Pearson C, 1990 Establishment of a conservator classification structure in Australia. In Stanley-Price N (ed), *The Graduate Conservator in Employment, Expectations and Realities. Proceedings of the Interim Meeting of the Working Group on Training in Conservation and Restoration of the ICOM Committee of Conservation, Amsterdam and Leiden 31 August-1 September 1989*, Amsterdam, Oplieding Restauratoren, 123–136.

Pearson C, 1999 Competency standards for conservation and their assessment: are they applicable to conservation training? In Dardes K (ed), *Defining and Measuring Effectiveness in Education and Training. Proceedings of the Interim Meeting of the ICOM-CC Working Group on Training in Conservation and Restoration, April 1998, Vantaa, Finland*, Vantaa, Evtek Institute of Art and Design, 77–96.

Pearson C and Ferguson R, 1996 A proposal for guidelines for conservation training. In Cronyn J and Foley K (eds), *A Qualified Community, Towards Internationally Agreed Standards of Qualification for Conservation. Proceedings of the Interim Meeting of the Working Group on Training in Conservation and Restoration of the ICOM Committee for Conservation Maastricht, April 1995*, London, ICOM-CC Working Group on Training in Conservation and Restoration, 63–67.

Pearson C and Winsor P (eds), 1986 *Role of Science in Conservation Training. Proceedings of the Interim Meeting of the ICOM-CC Working Group on Training in Conservation and Restoration, October 1986, British Museum*, London, ICOM.

Peck W H, 1980 Mummies of Ancient Egypt. In Cockburn A, Cockburn E and Reyman T A (eds), *Mummies, Disease and Ancient Cultures*, Cambridge, Cambridge University Press, 15–37.

Peers L, 1999 Curating Native American art, *British Museum Magazine, The Journal of the British Museum Society* 34, 24–27.

Perier-d'Ieteren C, 1998 Tourism and conservation: striking a balance, *Museum International* 50(4), 5–14.

Perry R, 1990 Problems of dirt accumulation and its removal from unvarnished paintings. In Hackney S, *et al* (eds), *Dirt and Pictures Separated*, London, UKIC, 3–6.

Petrie W M F, 1888 The treatment of small antiquities, *Archaeological Journal* 45, 85–89.

Petzet M, 1995 'In the full richness of their authenticity': the test of authenticity and the new cult of monuments. In Larsen K E (ed), *Nara Conference on Authenticity in Relation to the World Heritage Convention, November 1994*, Tokyo, UNESCO World Heritage Centre; Agency for Cultural Affairs (Japan); ICCROM; ICOMOS, 85–100.

Philippot P, 1976 Historic preservation: philosophy, criteria, guidelines. In International Centre for Conservation, Rome, and International Centre Committee of the Advisory Council on Historic Preservation, *Preservation and Conservation, Principles and Practices. Proceedings of the North American International Regional Conference, 1972*, Washington DC, Preservation Press, 367–382.

Philippot P, 1996 Restoration from the perspective of the humanities. In Stanley-Price N, Talley M K and Vaccaro A M (eds), *Historical and Philosophical Issues in the Conservation of Cultural Heritage*, Los Angeles, The Getty Conservation Institute, 216–229.

Phillips D, 1997 *Exhibiting Authenticity*, Manchester and New York, Manchester University Press.

Pickford C, Rhys-Lewis J and Weber J, 1997 *Preservation and Conservation*, Best Practice Guideline 4, London, Society of Archivists.

Picton J (ed), 1984 *Earthenware in Asia and Africa.* Colloquies on Art and Archaeology in Asia 12, London, Percival David Foundation of Chinese Art.

Picton J and Mack J, 1979 *African Textiles*, London, British Museum Publications.

Pinelli O R, 1996 The surgery of memory: ancient sculpture and historical restorations. In Stanley-Price N, Talley M K and Vaccaro A M (eds), *Historical and Philosophical Issues in the Conservation of Cultural Heritage*, Los Angeles, The Getty Conservation Institute, 288–305.

Pinniger D, 1991 *Insect Pests in Museums*, London, Archetype.

Pinniger D and Winsor P, 1998 *Integrated Pest Management*, London, Museums and Galleries Commission.

Pitt Rivers Museum, 1997 Publicity leaflet for an exhibition entitled *Braving the Elements* (exhibition held from November 1997 to December 1998).

Plenderleith H J, 1934 *The Preservation of Antiquities*, London, The Museums Association.

Plenderleith H J, 1956 *The Conservation of Antiquities and Works of Art*, London, Oxford University Press.

Plenderleith H J, 1998 A history of conservation, *Studies in Conservation* 43, 129–143.

Pliny the Elder *Natural History*, trans 1938, Rackham H, Cambridge, MA, Harvard University Press.

Plowden A and Halahan F, 1987 *Looking after Antiques*, London and Sydney, Pan Books.

Podany J C, 1987 Advances in the reassembly of large stone sculpture at the J Paul Getty Museum. In Black J (ed), *Recent Advances in the Conservation and Analysis of Artifacts*, London, Summer Schools Press, 375–382.

Podany J C, 1994 Restoring what wasn't there: reconsideration of the eighteenth-century restorations to the Lansdowne Herakles in the collection of the J Paul Getty Museum. In Oddy A (ed), *Restoration, Is it Acceptable?*, British Museum Occasional Paper 99, London, British Museum, 9–18.

Podany J C, Elston M, Thoresen L and Maish S L, 1995 The use of paper-pulp-based fill material for compensation of losses to sculpture. In Heuman J (ed), *From Marble to Chocolate, the Conservation of Modern Sculpture, Tate Gallery Conference September 1995*, London, Archetype, 59–64.

Podany J C and Maish S L, 1993 Can the complex be made simple? Informing the public about conservation through museum exhibits, *Journal of the American Institute for Conservation* 32, 101–108.

Podany J C and Scott D, 1997 Looking through both sides of the lens: why scientists and conservators should know each other's business. In Bradley S (ed), *The Interface between Science and Conservation*, British Museum Occasional Paper 116, London, British Museum, 211–220.

Portell J D, 1992 Colored glazes on silver-gilded surfaces. In Hodges H W M, Mills J S and Smith P (eds), *Conservation of the Iberian and Latin American Cultural Heritage. Preprints of the Constributions to the Madrid Congress, 9-12 September 1992*, London, International Institute for Conservation, 116–118.

Pretty G L and Calder A, 1980 Mummification in Australia and Melanesia. In Cockburn A, Cockburn E and Reyman T A (eds), *Mummies, Disease and Ancient Cultures*, Cambridge, Cambridge University Press, 289–307.

Preusser F, 1991 Scientific and technical examination of the tomb of Queen Nefertari at Thebes. In Cather S (ed), *The Conservation of Wall Paintings*, Los Angeles, The Getty Conservation Institute, 1–12.

Price C A, 1992 Training for research conservation. In White R (ed), *Archaeological Conservation, Training and Employment*, London, UKIC, 18–19.

Price C A, 1996 *Stone Conservation: An Overview of Current Research,* Santa Monica, CA, Getty Conservation Institute.

Price C A, 2000a Following fashion: the ethics of archaeological conservation. In McManamon F P and Hatton A (eds), *Cultural Resource Management in Contemporary Society, Perspectives on managing and presenting the past,* London and New York, Routledge, 213–230.

Price C A (ed), 2000b *ECOS, Environmental Control of Salts*, European Commission, Protection and Conservation of European Cultural Heritage, Research Report.

Prown J D, 1993 The truth of material culture: history or fiction? In Lubar S and Kingery W D (eds), *History from Things – Essays on Material Culture*, Washington and London, Smithsonian Institution Press, 1–19.

Pullen D, 1999 Managing change – the conservation of plastic sculptures. In Heuman J (ed), *Material Matters: The Conservation of Modern Sculpture*, London, Tate Gallery Publishing, 100–107.

Pullen D and Deighton S, 1995 Barbara Hepworth: Conserving a lifetime's work. In Heuman J (ed), *From Marble to Chocolate, the Conservation of Modern Sculpture, Tate Gallery Conference September 1995* London, Archetype, 136–143.

Pye E, 1992 Conservation and storage: archaeological material. In Thompson J M A (ed), *The Manual of Curatorship*, second edition, London, Butterworth-Heinemann, 392–426.

Pye E, 1999 Improving the route from conservation education to professional practice. In Dardes K (ed), *Defining and Measuring Effectiveness in Education and Training. Proceedings of the interim meeting of the ICOM-CC Working Group on Training in Conservation and Restoration, Vantaa, Finland,* Vantaa, Evtek Institute of Art and Design, 97–102.

Pye E and Cronyn J, 1987 The archaeological conservator re-examined. In Black J W (ed), *Recent Advances in the Conservation and Analysis of Artifacts*, Institute of Archaeology, Jubilee Conservation Conference Papers, London, Summer Schools Press, 355–357.

Rahtz P (ed), 1974 *Rescue Archaeology*, London, Penguin Books.

Ramer B, 1989 *A Conservation Survey of Museum Collections in Scotland*, Edinburgh, Scottish Museums Council.

Ramsay Jolicoeur B and Wainwright I N M (eds), 1990 *Shared Responsibility: A Seminar for Curators and Conservators*, Ottawa, National Gallery of Canada.

Rathgen F, 1905 *The Preservation of Antiquities*, Auden G A and Auden H A (trans), Cambridge, Cambridge University Press.

Reedy C, 1991 The opening of consecrated Tibetan bronzes with interior contents: scholarly, conservation, and ethical considerations, *Journal of the American Institute for Conservation* 30, 130–34.

Reedy C, 1992 Religious and ethical issues in the study and conservation of Tibetan sculpture, *Journal of the American Institute for Conservation* 31, 41–50.

Rees-Jones S G, 1990 Early experiments in pigment analysis, *Studies in Conservation* 35, 93–101.

Reilly J A, 1991 Celluloid objects: their chemistry and preservation, *Journal of the American Institute for Conservation* 30, 145–162.

Renfrew C, 1986 Varna and the emergence of wealth in prehistoric Europe. In Appadurai A (ed), *The Social Life of Things, Commodities in Cultural Perspective*, Cambridge, Cambridge University Press, 141–168.

RIBA (Royal Institute of British Architects), 1865 *Conservation of Ancient Monuments and Remains, General Advice to Promoters, Hints to Workmen*, London, RIBA.

Richardson J, 1996 Crimes against the Cubists. In Stanley-Price N, Talley M K and Vaccaro A M (eds), *Historical and Philosophical Issues in the Conservation of Cultural Heritage*, Los Angeles, The Getty Conservation Institute, 185–192.

Riegl A, 1996 The modern cult of monuments: its essence and its development. In Stanley-Price N, Talley M K and Vaccaro A M (eds), *Historical and Philosophical Issues in the Conservation of Cultural Heritage* Los Angeles, The Getty Conservation Institute, 69–83.

Robinson J and Pardoe T, 2000 *An Illustrated Guide to the Care of Costume and Textile Collections*, London, Museums and Galleries Commission.

Rolfe M, 1999 *Little Dancer Aged Fourteen* 1880-1, cast *c* 1922. In Heuman J (ed), *Material Matters, The Conservation of Modern Sculpture*, London, Tate Gallery Publishing, 14–25.

Rose C L and de Torres A R (eds), 1992 *Storage of Natural History Collections, Ideas and Practical Solutions*, Pittsburgh, Society for the Preservation of Natural History Collections.

Rose J, 1998 An investigation into the domestic care of paintings in English country houses in the eighteenth and nineteenth centuries. In Sitwell C and Staniforth S (eds), *Studies in the History of Painting Restoration*, London, Archetype Publications in association with the National Trust, 139–180.

Roy A, 1998 Forbes Prize lecture, *International Institute for Conservation [IIC] Bulletin* 6, 1–5.

Roy G, 1997 The relevance of science training to the conservation practitioner now. In Bradley S (ed), *The Interface between Science and Conservation*, British Museum Occasional Paper 116, London, British Museum, 25–30.

Rule M, 1982 *The Mary Rose*, London, Conway Maritime Press.

Ruskin J, 1883 *The Seven Lamps of Architecture*, fourth edition, Orpington, Kent, George Allen.

Samajdar S, 1997 Machine vision as a materials conservation technology. In Vandiver P B, Drusik J, Merkel J and Stewart J (eds), *Materials Issues in Art and Archaeology V*, Pittsburgh, Pennsylvania, Materials Research Society, 399–404.

Samet W, Hill Stoner J and Wolbers R, 1990 Approaching the cleaning of Whistler's 'Peacock Room': retrieving surface interrelationships in 'Harmony in Blue and Gold'. In Mills J S and Smith P (eds), *Cleaning, Retouching and Coatings, Technology and Practice for Easel Paintings and Polychrome Sculpture. Preprints of the Contributions to the Brussels Congress*, London, IIC, 6–12.

Samuel R, 1994 *Theatres of Memory, Vol 1, Past and Present in Contemporary Culture*, London, Verso.

Sanai D, 1999 Have I got ancient news for you, *The Independent*, 13th April.

Sandwith H and Stainton S, 1991 *The National Trust Manual of House-Keeping*, London, Penguin Books & National Trust.

Sayre E, 1993 Examination and analysis of objects in the museum laboratory. In Vandiver P B, Druzik J R, Freestone I C and Wheeler G S (eds), *Materials Issues in Art and Archaeology III*, Pittsburgh, Pennsylvania, Materials Research Society, xlv–xlvii.

Schleicher B, 1997 Problems in the cleaning of polychromed wood sculpture. In Lindley P (ed), *Sculpture Conservation, Preservation or Interference?*, Aldershot, Scolar Press, 151–154.

Schmidt H G, 1983 Problem-based learning: rationale and description, *Medical Education* 17, 11–16.

Schnapp A, 1996 *The Discovery of the Past, the Origins of Archaeology*, London, British Museum Press.

Schreiner M, 1987 Analytical investigations of medieval glass paintings. In Black J (ed), *Recent Advances in the Conservation and Analysis of Artifacts*, London, Summer Schools Press, 73–80.

Schultz R L, Clark D E and Folz D C, 1992 Theories of corrosion as applied to preservation of archaeological ceramics. In Vandiver P B, Druzik J R, Wheeler G S and Freestone I C (eds), *Materials Issues in Art and Archaeology III*, Pittsburgh, Pennsylvania, Materials Research Society, 1015–1030.

Scott A, 1926 *The Cleaning and Restoration of Museum Exhibits*, London, HM Stationery Office.

Scott D A, 1994 An examination of the patina and corrosion morphology of some Roman bronzes, *Journal of the American Institute for Conservation* 33, 1–23.

Scott G, 1994 Moisture, ventilation and mould growth. In Roy A and Smith P (eds), *Preventive Conservation, Practice Theory and Research. Preprints of the Contributions to the Ottawa Congress September 1994*, London, IIC, 149–153.

Sease C, 1981 The case against using soluble nylon in conservation work, *Studies in Conservation* 26, 102–110.

Sease C, 1987 (1994) *A Conservation Manual for the Field Archaeologist*, third edition, Los Angeles, Institute of Archaeology UCLA.

Sease C, 1996 A short history of archaeological conservation. In Roy A and Smith P (eds), *Archaeological Conservation and its Consequences, Preprints of the Contributions to the Copenhagen Congress*, London, IIC, 157–161.

Sease C, 1997 Conservation and the antiquities trade, *Journal of the American Institute for Conservation* 36, 49–58.

Sease C, 1998 Codes of ethics for conservation, *International Journal of Cultural Property* 7, 98–114.

Seeley N, 1987 Archaeological conservation: the development of a discipline, *Institute of Archaeology Bulletin* 24, 161–175.

Seeley N, 1999 Reversibility – achievable goal or illusion? In Oddy A and Carroll S (eds), *Reversibility – Does it Exist?*, British Museum Occasional Paper 135, London, British Museum, 161–168.

Seguel R, 1990 Development of a training program for professional restorers and design of an academic-professional profile in Chile. In Stanley-Price N (ed), *The Graduate Conservator in Employment, Expectations and Realities. Proceedings of the Interim Meeting of the Working Group on Training in Conservation and Restoration of the ICOM Committee of Conservation, Amsterdam and Leiden 31 August – 1 September 1989*, Amsterdam, Oplieding Restauratoren, 24–39.

Semple-Kerr J S, 1996 *The Conservation Plan, A Guide to the Preparation of Conservation Plans for Places of European Cultural Significance*, fourth edition, Sydney, The National Trust for Australia.

Sewell B, 2000 No deceit, no conspiracy, just a Rubens through and through, *London Evening Standard*, 2nd June.

Shanks M, 1992 *Experiencing the Past*, London and New York, Routledge.

Shanks M and Hodder I, 1995 Processual, postprocessual and interpretive archaeologies. In Hodder I, Shanks M, Alexandri A, Buchli V, Carman J, Last J and Lucas G (eds), *Interpreting Archaeology, Finding Meaning in the Past*, London and New York, Routledge, 3–29.

Shashoua Y, 1997 Leave it to the experts. In Bradley S (ed), *The Interface between Science and Conservation*, British Museum Occasional Paper 116, London, British Museum, 231–236.

Shearer G L and Doyal S, 1991 Use of FT-IR in the conservation of twentieth-century objects. In Vandiver P B, Drusik J and Wheeler G S (eds), *Materials Issues in Art and Archaeology II*, Pittsburgh, Pennsylvania, Materials Research Society, 813–824.

Sheldon R, 1988 *Management of the Collections of the English National Museums and Galleries. Committee of Public Accounts First Report*, London, Her Majesty's Stationery Office.

Shorer P, 1997 Observations of fifty years conservation at the British Museum. In MacLeod I D, Pennec S L and Robbiola L (eds), *Metal 95, Proceedings of the International Conference on Metals Conservation Semur en Auxois 25-28 September 1995*, London, James & James, 6–9.

Sickels-Taves L B, Sheehan M S and White G D, 1997 Applications of qualitative trace element analyses: an interdisciplinary approach to materials conservation. In Vandiver P B, Drusik J, Merkel J and Stewart J (eds), *Materials Issues in Art and Archaeology V*, Pittsburgh, Pennsylvania, Materials Research Society, 323–330.

Simpson M G, 1996 *Making Representations*, London and New York, Routledge.

Singer C, Holmyard E J and Hall A R, 1954–8 *A History of Technology* (five volumes), London, Oxford University Press.

Sitwell C, 1998 Approach to restoration in English country houses. In Sitwell C and Staniforth S (eds), *Studies in the History of Painting Restoration*, London, Archetype Publications in association with the National Trust, 129–138.

Smith B, 1990 Shared responsibility: welcome and introduction. In Ramsay-Jolicoeur B A and Wainwright I N M (eds), *Shared Responsibility, Proceedings of a Seminar for Curators and Conservators, National Gallery of Canada, 26th-28th October 1998*, Ottawa, National Gallery of Canada, 1–5.

Smith S, 1992 The Portland Vase. In Oddy A (ed), *The Art of the Conservator*, London, British Museum Press, 42–58.

Smith S, 1994 Filling and painting of ceramics for exhibition in the British Museum – is it acceptable? In Oddy A (ed), *Restoration, Is it Acceptable?*, British Museum Occasional Paper 99, London, British Museum, 159–165.

Smith S, 1999 Opacity contrariwise: the reversibility of deteriorated surfaces on vessel glass. In Oddy A and Carroll S (eds), *Reversibility – Does it Exist?*, British Museum Occasional Paper 135 London, British Museum, 135–140.

Sola T, 1992 Professionals and museums 1. In Boylan P J (ed), *Museums 2000, Politics, People, Professionals and Profit*, London and New York, Museums Association in conjunction with Routledge, 101–122.

SOS, 1998 *Save Outdoor Sculpture, Update* 9(1).

SPAB (Society for the Protection of Ancient Buildings), 1903 *Notes on the Repair of Ancient Buildings*, London, SPAB.

Spalding J, 1999 Creative management in museums. In Moore K (ed), *Management in Museums*, London and New Brunswick, Athlone Press, 28–44.

Specht J and MacLulich C, 1996 Changes and challenges: the Australian Museum and indigenous communities. In McManus P M (ed), *Archaeological Displays and the Public*, London, Institute of Archaeology University College London, 27–49.

Spindler K, Wilfing H, Rastbichler-Zissernig E, zur Nedden D and Nothdurfter H (eds), 1996 *Human Mummies, A Global Survey of their Status and the Techniques of Conservation. The Man in the Ice, vol 3*, New York, Springer Wien.

Spon, 1909 *Workshop Receipts. New and thoroughly revised edition*, London, E & F Spon.

St Clair W, 1999 The Elgin marbles: questions of stewardship and accountability, *International Journal of Cultural Property* 8, 391–521.

Staniforth S, 1990 Benefits versus costs in environmental control. In Keene S (ed), *Managing Conservation. Papers given at a Conference held jointly by UKIC and the Museum of London*, London, UKIC, 28–30.

Staniforth S, 1992 Control and measurement of the environment. In Thompson J M A (ed), *Manual of Curatorship*, London, Butterworth-Heinemann, 234–245.

Staniforth S, Hayes B and Bullock L, 1994 Appropriate technologies for relative humidity control for museum collections housed in historic buildings. In Roy A and Smith P (eds), *Preventive Conservation, Practice Theory and Research. Preprints of the Contributions to the Ottawa Congress September 1994*, London, IIC, 123–128.

Stanley-Price N, 1989 Archaeology and conservation training at the international level. In Cleere H (ed), *Archaeological Heritage Management in the Modern World*, London, Unwin Hyman, 292–301.

Stanley-Price N, 1990a Foreword. In Stanley-Price N (ed), *The Graduate Conservator in Employment: Expectations and Realities.* Proceedings of the Interim Meeting of the Working Group on Training in Conservation and Restoration of the ICOM Committee of Conservation, Amsterdam and Leiden 31 August–1 September 1989, Amsterdam, Oplieding Restauratoren, v–vi.

Stanley-Price N, 1990b The graduate conservator in employment, education in a changing discipline. In Stanley-Price N (ed), *The Graduate Conservator in Employment, Expectations and Realities. Proceedings of the Interim Meeting of the Working Group on Training in Conservation and Restoration of the ICOM Committee of Conservation, Amsterdam and Leiden 31 August – 1 September 1989*, Amsterdam, Oplieding Restauratoren, 137–148.

Stanley-Price N, Talley M K and Vaccaro A M (eds), 1996 *Historical and Philosophical Issues in the Conservation of Cultural Heritage*, Los Angeles, The Getty Conservation Institute.

Starzecka D C, 1996 *Maori Art and Culture*, London, British Museum Press.

Stone P, 1989 Interpretations and uses of the past in modern Britain and Europe. In Layton R (ed), *Who Needs the Past?*, London, Unwin Hyman, 195–206.

Stone T, 1990 Copyright and conservation. In Ramsay-Jolicoeur B A and Wainwright I N M (eds), *Shared Responsibility, Proceedings of a Seminar for Curators and Conservators, National Gallery of Canada, 26th-28th October 1998*, Ottawa, National Gallery of Canada, 149–152.

Stone T, 1996 Artifacts revisited: the evaluation of old treatments. In *ICOM Committee for Conservation, Preprints of the 11th Triennial Meeting, Edinburgh, September 1996*, London, James & James, 643–649.

Strong R, 1999 Obituary for Sir Hugh Casson, *London Evening Standard*, 17 August.

Strong R, Binney M and Harris J, 1975 *The Destruction of the Country House, 1875–1975*, London, Thames and Hudson.

Sturt G, 1923 (1930) *The Wheelwright's Shop*, second edition, Cambridge, Cambridge University Press.

Suenson-Taylor K, Sully D and Orton C, 1999 Data in conservation; the missing link in the process, *Studies in Conservation* 44(3), 184–194.

Sully D, 1992 Humidification: the reshaping of leather, skin, and gut objects for display. In Hallebeek P, Kite M and Calnan C (eds), *Conservation of Leathercraft and Related Objects. Interim Symposium*, London, Victoria and Albert Museum, 50–54.

Sully D and Suenson-Taylor K, 1996 A condition survey of glycerol treated freeze-dried leather in long-term storage. In Roy A and Smith P (eds), *Archaeological Conservation and its Consequences, Preprints of the Contributions to the Copenhagen Congress*, London, IIC, 177–181.

Summers J, 1995 Gilding the lily: the patination of Henry Moore's bronze sculpture. In Heuman J (ed), *From Marble to Chocolate, the Conservation of Modern Sculpture. Tate Gallery Conference September 1995*, London, Archetype, 144–151.

Swain H, 1998 *A Survey of Archaeological Archives in England*, London, Museums and Galleries Commission.

Swain H and Roberts M (eds), 1999 *The Spitalfields Roman*, London, Museum of London.

Talley M K, 1990 The humanistic foundation in the training of restorers. In Stanley-Price N (ed), *The Graduate Conservator in Employment, Expectations and Realities. Proceedings of the Interim Meeting of the Working Group on Training in Conservation and Restoration of the ICOM Committee of Conservation, Amsterdam and Leiden 31 August – 1 September 1989*, Amsterdam, Oplieding Restauratoren, 54–69.

Talley M K, 1992 Conservation II: 'A gyre and gimble in the wabe with the MTI's slithy toves', conservation qualifications and the Museum Training Institute, *International Journal of Museum Managment and Curatorship* 11(3), 317–320.

Talley, M K, 1998 Miscreants and hotentots: restorers and restoration attitudes and practices in seventeenth and eighteenth century England In Sitwell, C, and Staniforth, S, (eds) *Studies in the History of Painting Restoration* London, Archetype Publications in association with the National Trust, 27–42.

Talley M K, 1999 The Delta Plan: a nationwide rescue operation, *Museum International* 51, 11–15.

Taylor I, 1997 *Developing Learning in Professional Education*, Buckingham, Society for Research into Higher Education and Open University Press.

Taylor I and Burgess H, 1995 Orientation to self-directed learning: paradox or paradigm?, *Studies in Higher Education* 20, 87–98.

Tennent N H (ed), 1993 *Conservation Science in the UK. Preprints of the Meeting held in Glasgow May 1993*, London, James & James.

Tennent N H, 1997 Conservation science: a view from four perspectives. In Bradley S (ed), *The Interface between Science and Conservation*, British Museum Occasional Paper 116, London, The British Museum, 15–24.

Tennent N H, Baird T and Gibson L, 1996 The technical examination and conservation of blackened Delftware from anaerobic sites. In Roy A and Smith P (eds), *Archaeological Conservation and its Consequences, Preprints of the Contributions to the Copenhagen Congress*, London, IIC, 182–187.

Tennent N H, Tate J and Cannon L, 1993 The corrosion of lead artifacts in wooden storage cabinets, *Scottish Society for Conservation and Restoration Journal* 4, 8–10.

Theophilus, *On Divers Arts*, trans. 1963 Hawthorne J G and Smith C S. Chicago, University of Chicago Press.

Thickett D, 1998 Sealing of MDF to prevent corrosive emissions, *The Conservator* 22, 49–56.

Thomas R, 1999 Making sense of misery, *Museums Journal* January, 30–31.

Thompson J M A (ed), 1992 *The Manual of Curatorship*, Oxford, Butterworth-Heinemann.

Thomson G (ed), 1963 *Recent Advances in Conservation*, London, Butterworths, 211–214.

Thomson G (ed), 1968 *Contributions to the London Conference on Museum Climatology*, London, IIC.

Thomson G, 1986 *The Museum Environment*, London, Butterworth.

Thomson R, 1981 Tanning – man's first manufacturing process?, *Transactions of the Newcomen Society* 53, 139–155.

Thornes R 1995 *Protecting Cultural Objects through International Documentation Standards, A Preliminary Survey*, Santa Monica, The Getty Art History Information Program.

Thornes R, 1997 *Protecting Cultural Objects in the Global Information Society, The Making of Object ID*, Santa Monica, The Getty Art History Information Program.

Thornton J, 1998 A brief history and review of the early practice and materials of gap-filling in the West, *Journal of the American Institute for Conservation* 37, 3–22.

Tilley C (ed), 1993 *Interpretative Archaeology*, Oxford, Berg.

Tite M, 1972 *Methods of Physical Examination in Archaeology*, London, Seminar Press.

Tite M and Bimson M, 1987 Identification of early vitreous materials. In Black J (ed), *Recent Advances in the Conservation and Analysis of Artifacts*, London, Summer Schools Press, 81–86.

Torraca G, 1996 The scientist's role in historic preservation with particular reference to stone conservation. In Stanley-Price N, Talley M K and Vaccaro A M (eds), *Historical and Philosophical Issues in the Conservation of Cultural Heritage*, Los Angeles, The Getty Conservation Institute, 439–444.

Townley P and Parris R, 1994 *Caring for Heritage Objects*, Powerhouse Museum Research Series 3, New South Wales, Powerhouse Publishing.

Tubb K W (ed), 1995 *Antiquities Trade or Betrayed: Legal, Ethical and Conservation Issues*, London, Archetype Publications.

Tubb K W and Grissom C A, 1995 Ayn Ghazal, a comparative study of the 1983 and 1985 statuary caches, *Studies in the History and Archaeology of Jordan* V (5), 437–447.

Tubb K W and Sease C, 1996 Sacrificing the wood for the trees - should conservation have a role in the antiquities trade? In Roy A and Smith P (eds), *Archaeological Conservation and its Consequences, Preprints of the Contributions to the Copenhagen Congress*, London, IIC, 193–197.

Tylecote R F, 1962 *Metallurgy in Archaeology*, London, Edward Arnold.

Ucko P, 1989 Foreword. In Layton R (ed), *Conflict in the Archaeology of Living Traditions*, London and New York, Routledge, ix–xvii.

Ucko P J, 1992 Introduction. *World Archaeological Bulletin* 6, i–iii.

Ucko P (ed), 1995 *Theory in Archaeology*, London, Routledge.

Ucko P J, 2000 Enlivening a dead past, *Conservation and Management of Archaeological Sites* 4, 67–92.

UKIC (United Kingdom Institute for Conservation of Historic and Artistic Works), 1990 *Guidance for Archaeological Conservation Practice*, London, UKIC (http://www.netlink.co.uk/users/ukic/gacp.html) (accessed 12 August 1999).

UKIC (United Kingdom Institute for Conservation of Historic and Artistic Works), 1999 Code of Ethics and Rules of Practice. *Members Handbook*, London, UKIC.

UKIC Archaeology Section, 1982 *Excavated Artifacts for Publication, UK Sites. Guidelines* 1, London, UKIC.

Umney N, 1995 Documentation as a tool in the conservation of museum collections, *International Council of Museums – Conservation Committee [ICOM-CC] Study Series* 1, 23–25.

UNESCO, 1970 *Convention on the Means of Prohibiting and Preventing the Illicit Import, Export and Transfer of Ownership of Cultural Property,* Adopted by the General Conference, Paris 14 November 1970.

UNESCO, 1972 *Convention Concerning the Protection of World Cultural and Natural Heritage* Paris, UNESCO (http://www.unesco.org:80/whc).

UNIDROIT, 1995 *Unidroit Convention on Stolen or Illegally Exported Cultural Objects.*

Urzi C and Krumbein W E, 1994 Microbial impacts on the cultural heritage. In Krumbein W E, Brimblecombe P, Cosgrove D E and Staniforth S (eds), *Durability and Change. The Science, Responsibility, and Cost of Sustaining Cultural Heritage. Report of the Dahlem Workshop, December 1992,* Chichester, John Wiley, 107–135.

Usami N, 1988 The construction and repair of Japanese folding screens. In Mills J S, Smith P and Yamasaki K (eds), *The Conservation of Far Eastern Art. Preprints of the Contributions to the Kyoto Congress, September 1988,* London, IIC, 59–63.

Vandiver P, Druzik J R, Merkel J F and Stewart J (eds) 1997 *Materials Issues in Art and Archaeology V,* Pittsburgh, Materials Research Society.

Vandiver P and Wheeler G S, 1991 Introduction. In *Materials Issues in Art and Archaeology II,* Pittsburgh, Materials Research Society, xvii–xxvi.

Vaughan G R, 1990 The Piranesi vase. In Jones M (ed), *Fake? The Art of Deception,* London, British Museum Publications, 133.

Veliz Z, 1998 The restoration of paintings in the Spanish royal collections, 1734–1820. In Sitwell C and Staniforth S (eds), *Studies in the History of Painting Restoration,* London, Archetype Publications in association with The National Trust, 43–62.

Vickers M, 1995 Surface colour transfer from metal, ivory and stone to ceramic and glass. In *Materials Issues in Art and Archaeology IV,* Pittsburgh, Materials Research Society, 189–199.

Viollet-le-Duc E-E, 1996 Restoration. In Stanley-Price N, Talley M K and Vaccaro A M (eds), *Historical and Philosophical Issues in the Conservation of Cultural Heritage,* Los Angeles, The Getty Conservation Institute, 314–318.

Vitelli K D (ed), 1996 *Archaeological Ethics* Walnut Creek Ca, Altamira Press.

Vitruvius *The Ten Books of Architecture,* trans. 1960, Hicky Morgan M, New York, Dover Publications.

WAC, 1991 World Archaeological Congress First Code of Ethics, *World Archaeological Bulletin* 5, 22–23.

Wadum J and Noble P, 1999 Is there an ethical problem after the twenty-third treatment of Rembrandt's *Anatomy Lesson of Dr Nicholaes Tulp*? In Bridgland J (ed), *ICOM Committee for Conservation Preprints of the 12th Triennial Meeting Lyon 29 August – 3 September 1999,* London, James & James, 206–210.

Wainwrigh, G, 1989 Saving the Rose, *Antiquity* 63, 430–435.

Walden S, 1985 *The Ravished Image,* London, Wiedenfeld and Nicholson.

Walker E and Wright A E, 1996 Medical education begins in the first year: problem-based, community-oriented teaching in a pre-clinical curriculum, *Academic Development* 2, 17–29.

Waller R, 1994 Conservation risk assessment: a strategy for managing resources for preventive conservation. In Roy A and Smith P (eds), *Preventive Conservation, Practice Theory and Research. Preprints of the Contributions to the Ottawa Congress, September 1994,* London, IIC, 12–16.

Wallert A, 1996 Deliquescence and recrystallization of salts in the Dead Sea Scrolls, In Roy A and Smith P (eds), *Archaeological Conservation and its Consequences, Preprints of the Contributions to the Copenhagen Congress,* London, IIC, 198–202.

Wallis J, 1997 Crawling from the wreckage, *Museums Journal* November, 26–27.

Walmsley D and Buxton K, 1999 *Whitby Headland Excavations, Information for Teachers,* London, English Heritage Education Service.

Walter N, 1999 It's art, but is it worth it?, *The Independent,* 5 July.

Ward P, 1986 *The Nature of Conservation, A Race against Time,* Marina del Rey, The Getty Conservation Institute.

Warren J, 1998 The historic context: principles and philosophies. In Warren J, Worthington J and Taylor S (eds), *Context, New Buildings in Historic Settings,* Oxford, Architecture Press, 8–17.

Watkins S and Cameron E, 1987 An Iron Age warrior's grave from Guernsey: the excavation, conservation and interpretation of its contents. In Black J (ed), *Recent Advances in the Conservation and Analysis of Artifacts,* London, Summer Schools Press, 51–58.

Watkinson D and Neal V, 1998 *First Aid for Finds,* third edition, London, Rescue – The British Archaeological Trust, and Archaeology Section of the UK Institute for Conservation, with the Museum of London.

Watson-Smyth K, 2000a Sculpture stolen by tomb raider returns to Egypt, *The Independent*, 22 January.

Watson-Smyth K, 2000b Adopt-a-book appeal to preserve British Library's crumbling pages, *The Independent*, 23 May.

Weil S, 1994 The ongoing pursuit of professional status. In Kavanagh G (ed), *Museum Provision and Professionalism*, London and New York, Routledge, 251–256.

Welsh E, Sease C, Rhodes B, Brown S and Clavir M, 1992 Multicultural participation in conservation decision-making, *WAAC (Western Association for Art Conservation) Newsletter* 14(1), 13–22.

van de Wetering E, 1996 The autonomy of restoration: ethical considerations in relation to artistic concepts. In Stanley-Price N, Talley M K and Vaccaro A M (eds), *Historical and Philosophical Issues in the Conservation of Cultural Heritage*, Los Angeles, The Getty Conservation Institute, 193–199.

Wharton G, Blank S D, Dean J C, 1995 Sweetness and blight: conservation of chocolate works of art. In Heuman J (ed), *From Marble to Chocolate, the Conservation of Modern Sculpture. Tate Gallery Conference September 1995*, London, Archetype, 162–170.

White B and Ganiaris H, 1998 Excavating and analysing human skeletons. In Werner A (ed), *London Bodies*, London, Museum of London, 14–21.

Whittington A, 1998 Time to take care of conservation, *Museums Journal* July, 47.

Wijesuriya G 2000 Conserving the Temple of the Tooth Relic, *Public Archaeology* 1(2), 99–108.

Wikström L and Schießl U, 1996 European Confederation of Conservator-Restorers Organisations: the professional guidelines. In Cronyn J and Foley K (eds), *A Qualified Community, Towards Internationally Agreed Standards of Qualification for Conservation*, London, ICOM-CC, 73–78.

Williams N, 1988 Ancient methods of repairing pottery and porcelain. In Daniels V (ed), *Early Advances in Conservation*, British Museum Occasional Paper 65, London, British Museum, 147–150.

Williams N, 1989 *The Breaking and Remaking of the Portland Vase*, London, British Museum.

Williams N, 1992 The Sutton Hoo helmet. In Oddy A, *The Art of the Conservator*, London, British Museum Press, 73–88.

Williams S L, 1997 Preventive conservation: the evolution of a museum ethic. In Edson G (ed), *Museum Ethics*, London and New York, Routledge, 198–206.

Wills B, 1995 Some methods of basketry repair, using Japanese tissue paper and starch paste. In Hill L and Giles S (eds), *Where to Start, Where to Stop. Papers from the British Museum / MEG Ethnographic Conservation Colloquium*, Museum Ethnographers Group Occasional Papers 4, Hull, Museum Ethnographers Group, 109–114.

Wills B, Shashoua Y and Sully D, 1992 Approach to the conservation of a Mexican saddle and *anquera*. In Hodges H W M, Mills J S and Smith P (eds), *Conservation of the Iberian and Latin American Cultural Heritage. Preprints of the Contributions to the Madrid Congress, 9-12 September 1992*, London, International Institute for Conservation, 179–183.

Wilson A, Tuross N and Wachowiak M, 1996 Blood residues on archaeological objects – a conservation perspective. In Roy A and Smith P (eds), *Archaeological Conservation and its Consequences, Preprints of the Contributions to the Copenhagen Congress*, London, IIC, 213–218.

Wilson D, 1982 Public interest versus conservation, *Museum* 34(1), 65–67.

Wilson D, 1999 Here's mud in your eye, *The Independent*, 26th July.

Windell J and Nightingale J, 1999 Conservation by contract only, *Museums Journal* May, 15.

Winsor P, 1998 *Conservation and Collection Care Provision in the UK, Service Provision in the Institutional Sector*, London, Museums and Galleries Commission.

Winsor P, 2000 *Working with Independent Conservators*, London, MGC.

Wolfe S J and Mibach L, 1983 Ethical considerations in the conservation of Native American sacred objects, *Journal of the American Institute for Conservation* 23, 1–6.

Wright M and Player-Dahnsjö Y M T (eds), 1998 *Site Effects, The Impact of Location on Conservation Treatments*, Edinburgh, The Scottish Society for Conservation and Restoration.

Yates B, 1993 Coming to terms with change: the new curator, *Museum International* 45(4), 41–45.

Yates T and Davison S, 1996 Tile pavement and floor wear. In Teutonico J M (ed), *A Future for the Past*, London, English Heritage, 99–107.

Yoon Y H and Brimblecombe P, 2000 Contribution of dust at floor level to particle deposit within the Sainsbury Centre for the Visual Arts, *Studies in Conservation* 45, 127–137.

Young S, 1989 *'The Work of Angels' Masterpieces of Celtic Art, Sixth to Ninth Centuries AD*, London, British Museum Publications.

Yourcenar M, 1996 The mighty sculptor, Time. In Stanley-Price N, Talley M K and Vaccaro A M (eds), *Historical and Philosophical Issues in the Conservation of Cultural Heritage*, Los Angeles, The Getty Conservation Institute, 212–215.

Zhong Z B, 1988 The preservation of ancient Chinese paper. In Mills J S, Smith P and Yamasaki K (eds), *The Conservation of Far Eastern Art. Preprints of the Contributions to the Kyoto Congress, September 1988*, London, IIC, 19–21.

Index